Advances in Intelligent Systems and Computing

Volume 269

Series editor

Janusz Kacprzyk, Polish Academy of Sciences, Warsaw, Poland
e-mail: kacprzyk@ibspan.waw.pl

For further volumes:
http://www.springer.com/series/11156

About this Series

The series "Advances in Intelligent Systems and Computing" contains publications on theory, applications, and design methods of Intelligent Systems and Intelligent Computing. Virtually all disciplines such as engineering, natural sciences, computer and information science, ICT, economics, business, e-commerce, environment, healthcare, life science are covered. The list of topics spans all the areas of modern intelligent systems and computing.

The publications within "Advances in Intelligent Systems and Computing" are primarily textbooks and proceedings of important conferences, symposia and congresses. They cover significant recent developments in the field, both of a foundational and applicable character. An important characteristic feature of the series is the short publication time and world-wide distribution. This permits a rapid and broad dissemination of research results.

Yong-Tae Kim · Ichiro Kobayashi
Euntai Kim
Editors

Soft Computing in Advanced Robotics

Springer

Editors
Yong-Tae Kim
Hankyong National University
Gyeonggi-do
Korea

Euntai Kim
Yonsei University
Seoul
Korea

Ichiro Kobayashi
Ochanomizu University
Tokyo
Japan

ISSN 2194-5357 ISSN 2194-5365 (electronic)
ISBN 978-3-319-05572-5 ISBN 978-3-319-05573-2 (eBook)
DOI 10.1007/978-3-319-05573-2
Springer Cham Heidelberg New York Dordrecht London

Library of Congress Control Number: 2014933401

Printed on acid-free paper

Springer is part of Springer Science+Business Media (www.springer.com)

Preface

Intelligent system and robotics are inevitably bound up; intelligent robots makes embodiment of system integration by using the intelligent systems. We can figure out that intelligent systems are to cell units, while intelligent robots are to body components. The two technologies have been synchronized in progress. Making leverage of the robotics and intelligent systems, applications cover boundlessly the range from our daily life to space station; manufacturing, healthcare, environment, energy, education, personal assistance, logistics. This book aims at presenting the research results in relevance with intelligent robotics technology. We propose to researchers and practitioners some methods to advance the intelligent systems and apply them to advanced robotics technology. This book consists of 10 contributions that feature mobile robots, robot emotion, electric power steering, mulit-agent, fuzzy visual navigation, adaptive network-based fuzzy inference system, swarm EKF localization and inspection robot. This edition is published in original, peer reviewed contributions covering from initial design to final prototypes and authorization.

To help readers understand articles, we describe the short introduction of each article as follows;

1. "Network-Based Subsumption Architecture for Broadcast Control of Multiple Mobile Robots Based on a Poor Hardware/Software Platform": The article says, although each robot has an only poor hardware platform with a limitation of software development, the proposed wireless network-based subsumption architecture enables them to be broadcastly controlled and to be hierarchically implemented a high-level software architecture such as a schooling behavior.

2. "Expressions of Emotions of Koala Robot Based on Laban Movement Analysis": This paper describes the mechanism and the control system of a koala robot which is able to express various kinds of emotions based on the Laban Movement Analysis. In order to react to an action, the robot should have a lot of sensors such as a force sensor, a distance sensor, and an infrared sensor.

3. "Development of Steering Controller for Electric Vehicle": This development is that the exited On/Off signals calculated by the steering controller for the rotation speed and direction of the motor are transmitted to the stepping motor. A coupler is installed

between the central axes of the motor and the position sensor in order to protect the sensor.

4. "Design of Cruise Control System for Electric Vehicle Using Piece-Wised Control": This article proposes a design scheme of a speed controller for an unmanned ground vehicle's speed compensator using a piece-wised control. The performance of nonlinearity can be improved through the closed-loop control to compensate it.

5. "Vibration Minimization of Tower Typed 2-Wheeled Mobile Robot Using Acceleration and Deceleration Velocity Profile Method": This paper proposes the velocity profile method with acceleration and deceleration to minimize the vibration of the 2-wheeled mobile robot. They have experimented three axis gyro sensors and three axis acceleration sensors mounted on the top of robot to measure the vibration and to analyze the effects of velocity profile.

6. "A Multi-Agent Context-Based Personalized User Preference Profile Construction Approach": This paper proposes a context-based personalized user preference profile construction approach to comprehensively track the user's local behaviors and user's web behavior of new inputted query, so that user can avoid the limitation of different vector.

7. "Fuzzy Visual Navigation Method for Autonomous Freight Transportation Robot": This paper proposes a fuzzy visual navigation method for the freight transportation robot. While the robot is driving to carry out transportation function in the various environments, it uses sensor information from ultrasonic sensors, RFID sensors and camera. A virtual map based on RFID nodes is designed and a path is generated for navigation.

8. "Design of Adaptive Network-based Fuzzy Inference System for Obstacle Avoidance of Mobile Robot": This paper utilizes the ANFIS method to record the path to the destination and to provide more rules to evaluate the situation. It is then possible to change the output and rules of the fuzzy inference system to obtain a better result. It shows the feasibility of obstacle-avoidance performance using the ANFIS method.

9. "Swarm EKF Localization for a Multiple Robot System with Range-Only Measurements": This paper proposes swarm EKF localization, a hybrid of two inference algorithms, extended Kalman filter (EKF) and belief propagation (BP), with a capability of choosing how many dependencies of random variables are exploited in inference using the concept of neighborhood. Also, this paper presents a numerical experiment result of swarm EKF localizations.

10. "Inspection Robot for Parallel Entry Boiler Header Pipe": This article presents the development of inspection robot for parallel entry boiler header pipes with drastic change in diameter. The robot is able to retract when entering the parallel entry and expand in bigger pipe. Camera gimbal mechanism enables the robot to acquire 360° inner pipe surface image with its position.

We would appreciate it if readers could get useful information from the articles and contribute to creating innovative and novel technology or theory on the advanced robotics. Thank you.

Yong-Tae Kim
Ichiro Kobayashi
Euntai Kim

Contents

Network-Based Subsumption Architecture for Broadcast Control of Multiple Mobile Robots Based on a Poor Hardware/Software Platform

Fusaomi Nagata[1], Akimasa Otsuka[1], Keigo Watanabe[2], and Maki K. Habib[3]

[1] Tokyo University of Science, Yamaguchi 756-0884, Japan
nagata@rs.tus.ac.jp
http://www.yama.tus.ac.jp/
[2] Okayama University, Okayama 700-8530, Japan
[3] American University in Cairo, Cairo 11511, Egypt

Abstract. In this paper, a wireless network-based subsumption architecture is proposed for a broadcast control of multiple mobile robots and for reduction of the total design cost required for constructing the system. When designing an actual experimental system using multiple mobile robots, there exits a serious requirement to suppress the total design cost with the increase of the number of robots. Even though each robot has an only poor hardware platform with a limitation of software development, the proposed wireless network-based subsumption architecture enables them to be broadcastly controlled and to be hierarchically implemented a high-level software architecture such as a schooling behavior. Also, this system enables the multiple mobile robots to be omitted the complicated download process of hex code program in debugging work because the debugging is conducted only on a supervisory server, so that the high maintainability can be performed. The multi-robots monitor running on the supervisory server and the agent dispatcher determining to which agent the highest priority of execution should be given, are further described.

Keywords: multiple mobile robots, network-based subsumption architecture, broadcast control, server supervisory control, Multi-robots monitor, Agent dispatcher, PSD sensor, flex sensor.

1 Introduction

Many network-based robotic systems and multiple mobile robots systems have been proposed according to various objectives. First of all, let us survey some valid papers mainly with some actual experimental results in chronological order. Parsons and Canny proposed an algorithm for planning the motions of several mobile robots which share the same workspace containing polygonal obstacles [1]. Each robot has an ability of independent translational motion in two dimensions. The algorithm computes a path for each robot which avoids all obstacles in the workspace as well as the other robots. Barman et al. developed an extensible facility for multiple mobile robots [2]. The system consists of nine radio-controlled mobile robots, two CCD color video cameras, a video

Y.-T. Kim, I. Kobayashi, and E. Kim (eds.), *Soft Computing in Advanced Robotics*
Advances in Intelligent Systems and Computing 269,
DOI: 10.1007/978-3-319-05573-2_1,

transmitter and tuner, radio controllers, an image processing hardware and so on. Software for tracking control is described. Kube and Zhang examined the problem of controlling multiple behavior-based autonomous robots [3,4]. Based on observations made from the study of social insects, they proposed five simple mechanisms used to invoke group behavior in simple sensor-based mobile robots. They also constructed a system of five homogeneous sensor-based mobile robots with capability of achieving simple collective task. Noreils described an architecture for cooperative and autonomous mobile robots [5]. The cooperation is composed of two phases. One is the collaboration where a task is decomposed into subtasks. The other is the coordination where robots actually coordinate their activities to fulfill the initial task using the notion of coordinated protocols. This architecture showed benefits of modularity, robustness and programmability.

Also, Azarm and Schmidt presented a novel approach to do conflict-resolution for multiple mobile robots [6]. A framework for negotiation is developed by using the online motion planning, which permits quick decentralized and parallel decision-making. The key objective of the negotiation procedure is dynamic assignment of robot motion priorities. The basic performance is evaluated from experiments using only two mobile robots. Bennewitz and Burgard considered the problem of path planning for teams of mobile robots [7]. It presented a decoupled and prioritized approach to coordinate the movements of the mobile robots in their environment. The proposed algorithm computes the paths for the individual robots in the configuration-time space. To estimate the risk of colliding with other robots, it uses a probabilistic model of the robots motions. Guo and Parker proposed a distributed and optimal motion planning algorithm for multiple robots, in which the computationally expensive problem was decomposed into two modules, i.e., path planning and velocity planning [8]. The D* search method was applied in both modules, based on either geometric formulation or schedule formulation. The algorithm was implemented in a group of Nomad 200 indoor robots, so that successful results were obtained. Parker outlined the project that demonstrated a team of 100+ heterogeneous robots solving an indoor reconnaissance and surveillance task [9]. The specific problem to be solved was shown with the robot team. The focus was on the impact of heterogeneity on the collaborative solution approach that the robot team must take. Pimentel and Campos addressed the problem of multi-mobile robot cooperation with strict communication constraints which are considered indispensable for successful task execution [10]. The problem is instantiated as a cooperative search and rescue task, and is modeled as a minimization of an energy functional which accounts for network connectivity, other relevant robot and task requirements in order to select locally optimal actions for each robot.

Antonelli et al. presented two experimental case studies performed using a multi-robots system made of six Khepera II mobile robots [11]. The experiments are aimed at testing the performances and the robustness of a behavior-based technique, called the null-space-based behavioral control (NSB). The NSB approach was developed to control a generic team of autonomous vehicles and it was implemented on a centralized architecture to control a platoon of autonomous mobile robots at a kinematic level [12]. Also, Mondada et al. developed a particular designed mobile robot called e-puck which targets engineering education at university level [13].

As for network-based system, for example, Zhan et al. presented the autonomous mobile robot interactive behavior in ubiquitous computing environment, in which multimodal network-based interfaces for human-robot interactions (HRI) was constructed. The mobile robot consisted of seven systems including vision, speech, remote supervisory and sensory systems, locomotion, robotic arms and power management systems [14]. Han and Payandeh developed a novel low-cost micro mobile robotic system which could be used as a part of the research and development in sensor and networked-based dynamical system for indoor applications. The developed system could be integrated as a part of the sensor fusion for localization of mobile agents in complement with information from the camera network [15].

The subsumption architecture proposed by Brooks is one of typical behavior-based artificial intelligences [16]. The subsumption architecture has been successfully implemented in the controllers of various types of robots, where layers of control system are built to let the robot operate at increasing levels of competence. In this paper, a wireless network-based subsumption architecture is proposed for broadcast control of multiple mobile robots and for reduction of the total design cost required for constructing the system. When designing an actual experimental system, i.e., not simulation, by using multiple mobile robots, there exits a serious demand to suppress the total design cost that is proportional with the increase of the number of robots. Even though each robot has only simple hardware platform with limitations of software development environment, the proposed architecture enables to construct a high-level software architecture such as a schooling behavior. The schooling behavior means that the robots behave like carps and other fishes in a Japanese artificial circular pond. Multi-robots monitor running on the supervisory server and agent dispatcher determining to which agent the highest priority of execution should be given, are described. In addition, this system enables the multiple mobile robots to be removed the complicated download process of hex code program in debugging work because the debugging is conducted only on a supervisory server, so that the high maintainability can be achieved. The basic processing capability of the wireless network-based subsumption architecture is experimentally evaluated in association with the easy measurement of the communication time, which can be regarded as a metrics similar to round-trip time or latency in networking. A simple and reliable measurement method using a Windows multimedia timer and packet data is introduced.

2 Multiple Mobile Robots System

2.1 Basic Structure of Multiple Mobile Robots

Figure 1 shows the developed multiple mobile robots system, in which each of them equips with six distance sensors [17]. The main body of the robot is an omni-directional mobile robot with three wheels provided by TOSADENSHI LTD [18]. A MicroConverter ADuC814ARU provided by ANALOG DEVIVES is mounted on the control board of the mobile robot. A simple DC motor without an encoder is built in each wheel, so that the robot has a high cost performance. In order to measure the distances to objects in real time, the robot is equipped with six PSD (Position Sensitive Detector) sensors as shown in Fig. 2. $\boldsymbol{d}_i(k) = [d_{i1}(k)...d_{i6}(k)]^T$ is the distance vector of the i-th mobile robot at the discrete time k. The PSD sensor is mainly composed of LEDs, electrical resistances and photodiodes, and can calculate the distance to an object through triangulation technique. In order to cope with the problem of narrow directivity, the required number of the PSD sensor was increased to six. In fact, more than six PSD sensors may be desirable to further reduce the dead angle. Besides, each robot has a Bluetooth module to communicate with a server PC.

Fig. 1. Multiple mobile robots with three wheels and six PSD sensors

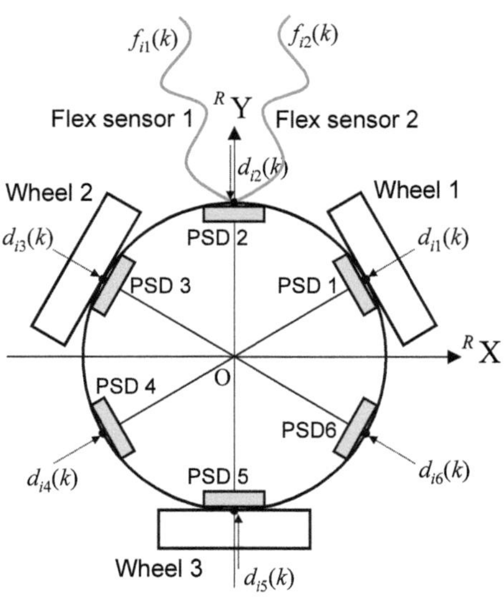

Fig. 2. Configuration of six PSD sensors and two flex sensors fixed around the mobile robot

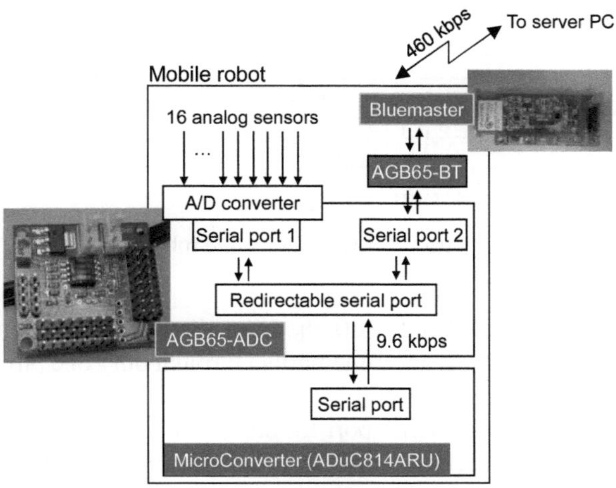

Fig. 3. Hardware block diagram of a mobile robot with a compact A/D conversion module AGB65-ADC and a Bluetooth module Bluemaster

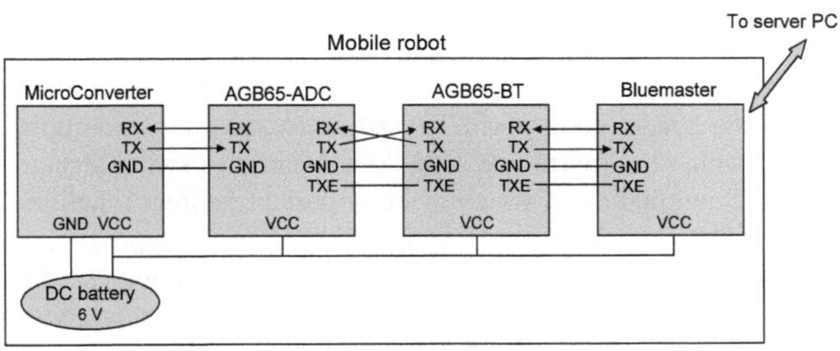

Fig. 4. Connection among the Microconverter, AGB65-ADC, and Bluemaster mounted on the mobile robot

2.2 Multiple Sensory Sensors

The mobile robot basically has six A/D conversion channels, however, all of which are already connected to the six PSD sensors. In order to further deal with sensory information from other sensors such as temperature, humidity, force and smell, a compact A/D conversion module AGB65-ADC provided by Asakusagiken Co., Ltd. [19] is mounted on the mobile robot. The AGB65-ADC is a high performance A/D converter that can further handle sixteen analogue channels. Allowed analog input range is from 0 to 5V, and an analog value is converted to either of two types of digital values, i.e., the maximum is 255 (8 bits) or 4095 (12 bits). In case of Fig. 2, two flex sensors are further added to detect objects within 10 cm. Figure 3 illustrates the hardware block diagram showing the connection scheme in the mobile robot including analog sensors, AGB65-ADC and a Bluetooth module called Bluemaster [20]. The AGB65-BT works as the signal level converter between AGB65-ADC [5V] and Bluemaster [3.3V]. Note that by

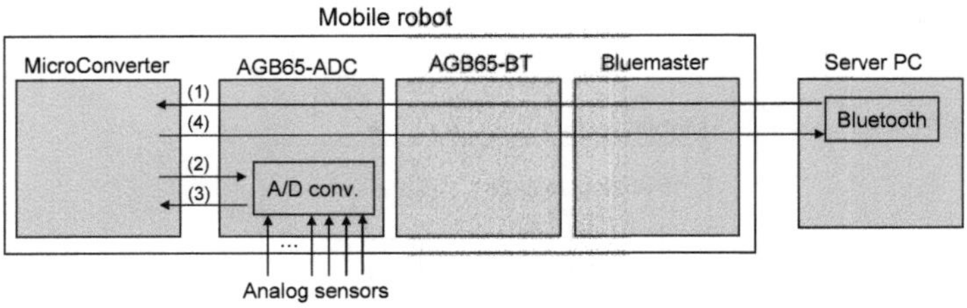

Fig. 5. Data flow of packet between a mobile robot and a server PC

using control codes, e.g., 0xFF and 0x00, the MicroConverter can communicate with two modlues (i.e., AGB65-ADC and Bluemaster) through only one physical serial line consisting RX, TX and GND.

The Microconverter has one serial port for data communication, so that text codes can be transmitted to and received from the Bluemaster through the AGB65-BT and serial port 2 in the AGB65-ADC. The Bluemaster is used to wirelessly communicate with a supervisory server PC. Figure 4 shows the designed connection scheme among the Microconverter, AGB65-ADC, and Bluemaster mounted on the mobile robot. It should be noted that when communicating with the A/D converter part in AGB65-ADC to obtain sensor's information through serial port 1, a command sequence including a hexadecimal header code "0xFF" has to be transmitted at first. On the other hand, text data without the header code "0xFF" is artfully sent to the serial port 2. In other words, the redirectable serial port in AGB65-ADC board can switch data to serial port 1 or to serial port 2 automatically, which is the important redirect function of AGB65-ADC. Using the A/D conversion module AGB65-ADC, various types of small and light weight sensory sensors can be further added to the mobile robot with six PSD sensors.

Figure 5 shows the data flow of a packet between a mobile robot and a server PC. First of all, a data packet including a reflex action command is transmitted from the server PC to a mobile robot through the route (1). After that, the mobile robot gives a request command for sensing to the A/D converter part through the route (2), and immediately obtains the information from analog sensors through the route (3). Then, the mobile robot returns a data packet including sensor information to the server PC through the route (4). The process consisting of the four routes from (1) to (4) can form the basic feedback control loop including wireless communication.

2.3 Kinematic Control of Three-Wheeled Mobile Robot

Next, the kinematic control method of the mobile robot is explained. Figure 6 illustrates the kinematic model of the mobile robot in robot coordinate system $\Sigma_R(\mathrm{O} - {}^R\mathrm{X}{}^R\mathrm{Y})$. ω_i $(i = 1, 2, 3)$ is the angular velocity of each wheel. Also, by using the radius r of the wheel, $v_i = r\omega_i$ is the forward velocity of each wheel. If the position and orientation vector of the robot is given by $[x_r\ y_r\ \phi_r]^T$, then the velocity in the robot coordinate

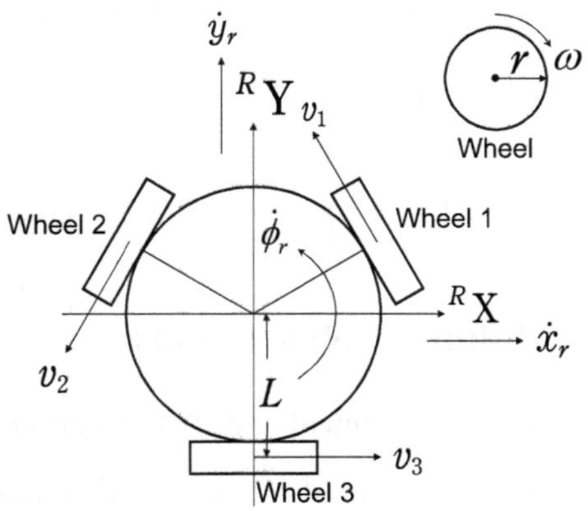

Fig. 6. Kinematics of the mobile robot with three wheels

system is represented by $[\dot{x}_r\ \dot{y}_r\ \dot{\phi}_r]^T$. For example, if it is assumed that the robot shown in Fig. 6 is now generating the velocity $[\dot{x}_r\ \dot{y}_r\ \dot{\phi}_r]^T$, then it contributes to the velocity v_1 in the robot coordinate system as

$$v_1 = -\frac{1}{2}\dot{x}_r + \frac{\sqrt{3}}{2}\dot{y}_r + L\dot{\phi}_r \tag{1}$$

In the same way, v_2 and v_3 can be represented by

$$v_2 = -\frac{1}{2}\dot{x}_r - \frac{\sqrt{3}}{2}\dot{y}_r + L\dot{\phi}_r \tag{2}$$

$$v_3 = \dot{x}_r + L\dot{\phi}_r \tag{3}$$

Consequently, the following kinematic relation is obtained [21].

$$\begin{pmatrix} \omega_1 \\ \omega_2 \\ \omega_3 \end{pmatrix} = \frac{1}{r}\begin{pmatrix} -\frac{1}{2} & \frac{\sqrt{3}}{2} & L \\ -\frac{1}{2} & -\frac{\sqrt{3}}{2} & L \\ 1 & 0 & L \end{pmatrix}\begin{pmatrix} \dot{x}_r \\ \dot{y}_r \\ \dot{\phi}_r \end{pmatrix} \tag{4}$$

where L is the distance between the center of the robot and the center of each wheel. By using Eq. (4), the robot can be controlled kinematically, i.e., desired behavior designed by $\dot{\boldsymbol{x}}_r = [\dot{x}_r\ \dot{y}_r\ \dot{\phi}_r]^T$ can be performed by making three wheels rotate with the angular velocity vector $\boldsymbol{\omega} = [\omega_1\ \omega_2\ \omega_3]^T$. As special cases, Table 1 shows the velocity components to move to the direction of each PSD sensor. When designing the schooling mode using multiple mobile robots, the six basic velocities tabulated in Table 1 are used. The important point is that the direction of velocity, which a mobile robot generates in Σ_R, depends on the ratio $\dot{x}_{ri} : \dot{y}_{ri}$. The velocity norm can be arbitrarily changed as $\alpha\boldsymbol{\omega} = [\alpha\omega_1\ \alpha\omega_2\ \alpha\omega_3]^T$ with a scalar α.

Table 1. Normalized velocity components in Σ_R to move to the direction of each PSD sensor

i	1	2	3	4	5	6
$\dot{x}_{ri}$	$\frac{\sqrt{3}}{2}$	0	$-\frac{\sqrt{3}}{2}$	$-\frac{\sqrt{3}}{2}$	0	$\frac{\sqrt{3}}{2}$
$\dot{y}_{ri}$	$\frac{1}{2}$	1	$\frac{1}{2}$	$-\frac{1}{2}$	-1	$-\frac{1}{2}$

3 Network-Based Multiple Mobile Robots

3.1 Software Development Environment and Its Limitation

The software development environment of the mobile robot using a free C language has two limitations with respect to functions. The first is, the flush ROM of the mobile robot is only 8 kB, so that it is difficult to construct a high level software architecture in the mobile robot. The second is, the mathematical standard library such as "exp( )" cannot be compiled. Thus, for example, it is impossible to directly apply the potential field method for path planning. In order to cope with such limitations of the software development environment, a server supervisory control scheme is considered. In the server control mode, each mobile robot integrates all information measured by six PSD sensors and transmits it to the server once, then the server can return a simple action command to the robot based on the sensor information while considering the overall behavior of a swarm. The mobile robots behave according to commands transmitted from the server. By means of the proposed server supervisory control, for example, the potential field method is available on the server side where the Windows Visual Studio runs as an environment for the software development, even though the mobile robot itself has only a simple software platform. This can be also applied to an actual experiment using multiple mobile robots where a software development with comparatively large scale is required.

3.2 Server Supervisory Control

In this subsection, a wireless network-based multiple mobile robots system using the server supervisory control is proposed in order to be able to design a high-level software architecture in spite of the limited capability of the robot software platform. The basic concept was briefly described in the past papers [20,22]. Figure 7 shows the conceptual diagram of the network-based multiple mobile robots system where the supervisory server can broadcastly interact only with selected multiple mobile robots. In this case, each robot can only transmit sensory information of distance and tactile to the supervisory server through Bluetooth communication. The distance and tactile can be measured with a PSD sensor and a flex sensor, respectively. The supervisory server returns a set of a simple behavior and a short execution time, e.g., 200 ms to the corresponding robot. Nine kinds of the most simply subdivided reaction behaviors, i.e., reflex actions, are prepared for the mobile robots as tabulated in Table 2. When a set of a command code and an execution time is transmitted from the supervisory server to a mobile robot, the mobile robot conducts the motion exactly within the specified execution time. Three agents called "Avoid objects", "Turn to left or right" and "Move forward" are designed by using the basic acts $A_i(k)$ $(i = 0, 1, 2, \cdots, 8)$ shown in Table 2 in order to realize a behavior like schooling of multiple robots.

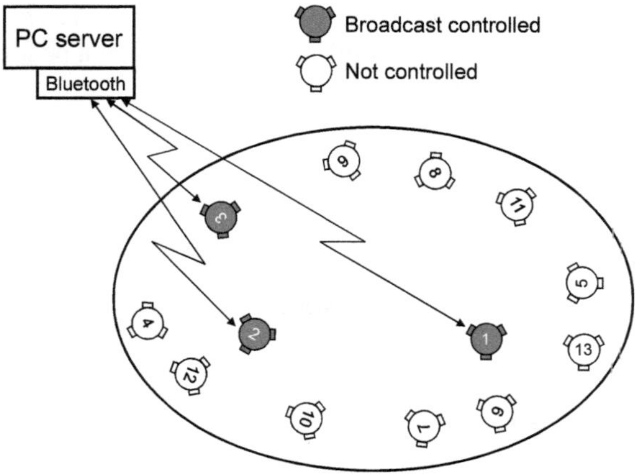

Fig. 7. Network-based multiple mobile robots system aiming at the emergence of a schooling behavior, where the supervisory server can broadcastly interact only with selected multiple mobile robots

Table 2. The most simply subdivided reaction behaviors, i.e., reflex actions, for mobile robots

Cmd. code	Corresponding motion
0	Halt at the position
1	Move to the direction of PSD sensor 1
2	Move to the direction of PSD sensor 2
3	Move to the direction of PSD sensor 3
4	Move to the direction of PSD sensor 4
5	Move to the direction of PSD sensor 5
6	Move to the direction of PSD sensor 6
7	Rotate to clockwise direction
8	Rotate to counterclockwise direction

3.3 Multi-robots Monitor

The proposed multi-robots monitor supervises the controller shown in Fig. 9. Figure 8 shows the software structure of the multi-robots monitor for N mobile robots. First of all, an arrival of packet from the first robot $(i = 1)$ is checked. If an arrival of packet is recognized, the packet is retrieved. Then, the agent dispatcher gives the execution right, i.e., the highest priority, to one of the agents according to sensor information included in the packet. The agent dispatcher does not immediately move the execution right to a higher priority agent when a lower priority agent is running, because each agent works as a simple reactive behavior every sampling period. Instead of this, whenever a reaction behavior is executed during a specified execution time, the dispatch monitor checks and updates the activity of each agent and gives the next execution right to a newly updated active agent with the highest priority. After that, the agent with the highest priority generates a command code denoting one of reflex acts tabulated in Table 2. Finally, a packet including the command code and an execution time is transmitted to the mobile robot. This process is applied to all mobile robot, i.e., until $i = N$. The proposed multi-robots monitor allows each of all the mobile robots to perform a behavior according to the designed subsumption architecture.

```
multi-robot_monitor( ) // Multimedia timer interrupt
{
   for(i = 0 ; i < N ; i++) // N : number of robots
   {
      if(packet_arrival(i) == TRUE)
      {
         packet_data = packet_receive(i) ;
         command_code = agent_dispatcher(packet_data) ;
         packet_send(i, command_code) ;
      }
   }
}
```

Fig. 8. Software structure of the proposed multi-robots dispatch monitor

4 Subsumption Architecture-Based Controller Implemented on Server PC

The software development environment is C# of Windows Visual Studio, which is used to develop and implement a high level software architecture such as subsumption control architecture according to application requirements. Figure 9 shows the networked-based subsumption architecture implemented on the supervisory server. M is the number of implemented agents. In experiment, the controller basically includes the three agents "Avoid objects", "Turn to left or right" and "Move forward" for a schooling behavior of multiple mobile robots. The upper agent has a higher priority to be dispatched. This section detailedly introduces the three kinds of agents and the corresponding output command codes shown in Table 2. The nine commands shown in Table 2 are simple and basic motions but important reflex actions for each mobile robot to consequently produce the competences of the three agents.

The server receives PSD sensors information $\boldsymbol{d}_i(k)(1 \leq i \leq N)$ from all mobile robots every sampling period, in which N is the number of the available mobile robots. By analyzing $\boldsymbol{d}_i(k)$, the controller dispatches the current execution right to one of the three agents for the i-th mobile robot. This process is periodically applied to all mobile robots in the order from 1 to N. In the schooling mode described in this section, all mobile robots try to regularly move along the inner of a circular fence keeping the distance to both the fence and other mobile robots. This mode enables the robots to behave like carps in a Japanese artificial circular pond.

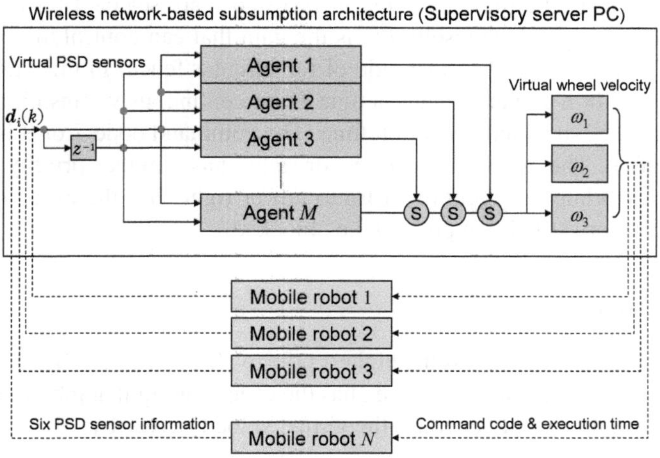

Fig. 9. Block diagram of network-based subsumption architecture for schooling behavior, which is implemented on supervisory server PC

4.1 "Move Forward"

When a robot moves counterclockwise along the inside of the circular fence as shown in Fig. 7, the agent of "Move forward" is always active. Accordingly, if higher priority agents such as "Turn to left or right" and "Avoid objects" are inactive, then the agent of "Move forward" can have the execution right and the following control law is applied.

$$\dot{\boldsymbol{x}}_i = v \frac{\dot{\boldsymbol{x}}_{i2}}{\|\dot{\boldsymbol{x}}_{i2}\|} \tag{5}$$

where $\dot{\boldsymbol{x}}_i = [\dot{x}_i \; \dot{y}_i]^T$ is the virtual translational velocity for the i-th mobile robot, $\dot{\boldsymbol{x}}_{i2} = [0 \; 1]^T$ is the normalized velocity vector for moving to the PSD sensor 2. The position of PSD sensor 2 is assumed to be the front of the mobile robot. v is the scalar signifying the magnitude of the velocity. While the agent of "Move forward" has the execution right, a set of a command code 2 and an execute time T, e.g., 200 ms, are continuously transmitted to the mobile robot every the specified execution time T.

4.2 "Turn to Left or Right"

The orientation of a mobile robot is controlled by the agent of "Turn to left or right". This agent becomes active when $d_{i1}(k) < d_{ref}$ and $d_{i6}(k) < d_{ref}$ are simultaneously satisfied. d_{ref} is called the active reference range which is like a threshold to be suitably set in advance. Further, if this agent has the execution right, then the following orientation control law is applied.

$$\dot{\phi}_i = K_\phi \{d_{i6}(k) - d_{i1}(k)\} \tag{6}$$

where $\dot{\phi}_i$ is the virtual rotational velocity of i-th mobile robot, $d_{i1}(k)$ and $d_{i6}(k)$ are the values of PSD sensors 1 and 6 transmitted from the i-th mobile robot, respectively.

As can be seen from the Fig. 2, PSD sensors 1 and 6 are selected for the orientation control moving counterclockwisely. K_ϕ is the gain that can control the orientation of the robot to be parallel to the inner side of the circular fence. In this case, a set of a command code 7 or 8 and an execution time T are continuously transmitted to the i-th mobile robot every specified execution time. The command code 7 or 8 is determined by the sign of $\dot{\phi}_i$. The agent "Turn to left or right" has a higher priority than "Move forward". Thus, while the agent of "Turn to left or right" has the execution right, the agent of "Move forward" is suppressed, i.e., blocked.

4.3 "Avoid Objects"

The agent of "Avoid objects" has the highest priority. If $\exists d_{ij}(k)\ [d_{ij}(k) < d_d]$ becomes true, then this agent becomes active, i.e., has the execution right at the same time. After becoming active, this agent generates the virtual velocity given by

$$\dot{\boldsymbol{x}}_i = -v\frac{\dot{\boldsymbol{x}}_{ij}}{\|\dot{\boldsymbol{x}}_{ij}\|}\{d_d - \min_j d_{ij}(k)\} \qquad (\exists d_{ij}(k)\ [d_{ij}(k) < d_d]) \tag{7}$$

where d_d is the minimum allowable distance between a robot and an object. The supervisory server transmits a set of a command code $j+3$ (in case of $j = 1, 2$ or 3) or $j-3$ (in case of $j = 4, 5$ or 6) and an execution time T to a mobile robot in order to avoid a collision with the nearest robot or object. Due to the activation of "Avoid objects" agent, the mobile robot can move away from moving objects including other mobile robots within the collision critical zone. If the distance to the nearest object is smaller than d_d, the robot tries to expand the distance to be over d_d. When multiple PSD sensors simultaneously detect shorter distances than d_d, the mobile robot tries to preferentially move away from the nearest object.

The finite state machine is illustrated in Fig. 10, which is a mathematical model of computation used to design both the computer program and the sequential logic. Figure 11 shows experimental scenes of schooling behavior, in which all mobile robots are broadcastly controlled based on the subsumption control architecture incorporated in the supervisory server as shown in Fig. 9. It was confirmed from the experiments that the multiple mobile robots could perform desirable schooling behaviors to with and without a large curvature fence.

4.4 "Slowly Turn to Left or Right"

"Avoid objects" actuates three wheels in order to avoid collisions to other robots within the range from 10 to 80 cm. For example, a higher priority agent "Slowly turn to left or right" can be easily designed using two flex sensors as shown in Fig. 2. This agent becomes active when the flex sensor detects an object within the near range, i.e., about 0 to 10 cm by checking the measurements $f_{i1}(k)$ and $f_{i2}(k)$. The control law of this agent is given by

$$\dot{\phi}_i = K_\phi\{f_{i2}(k) - f_{i1}(k)\} \tag{8}$$

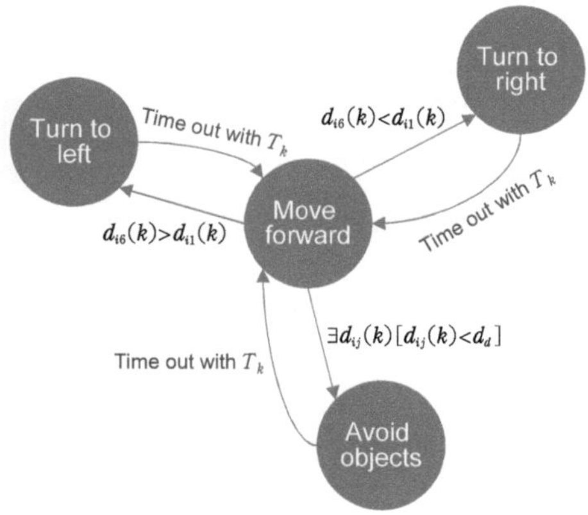

Fig. 10. Finite state machine among the three agents, "Move forward", "Turn to left/right" and "Avoid objects"

Fig. 11. Experiment of schooling behavior based on the proposed networked-based subsumption architecture, where all mobile robots are broadcastly controlled

By giving K_ϕ a smaller value than the case of Eq. (6), the robot rotates more slowly to the direction of less flexed sensor. In this case also, a set of a command code 7 or 8 and an execution time T are continuously transmitted to the i-th mobile robot every specified execution time.

As can be seen, a new agent can be easily added to the controller to produce an expectable competence.

4.5 Discussion of Communication Time

Communication time called the dynamic sampling period is an important factor to evaluate the performance of the network-based multiple mobile robots system [23]. The dynamic sampling period can be regarded as a metrics similar to round-trip time or latency in networking. The dynamic sampling period is defined for a server PC as variable time taking from submitting a command packet to a mobile robot to receiving a response packet including sensor information. Here, an easy measurement method of the dynamic sampling period is explained by using Windows multimedia timer. Figure 12 shows the handshake process to measure the dynamic sampling period $T_k = t_{k+1} - t_k$.

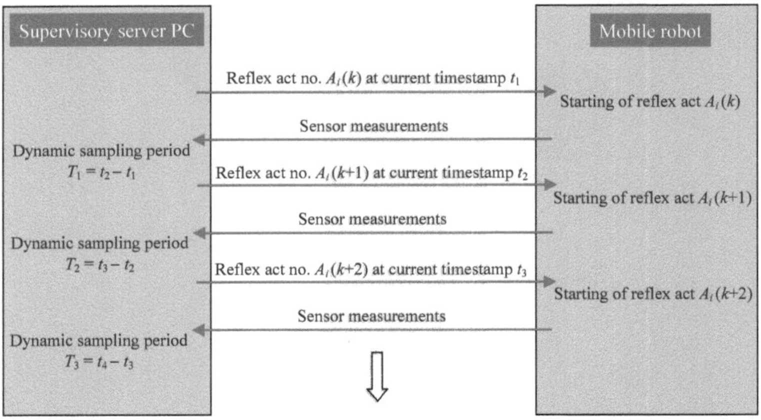

Fig. 12. Handshake process to measure the dynamic sampling period T_k by using Windows multimedia timer of 1 ms

The processor time t_k at the discrete time k in the server PC is used as a timestamp to synchronize the time between the server PC and each mobile robot. The processor time t_k can be obtained by counting up a counter variable in a Windows multimedia timer interrupt at the server PC side. The sampling period of the Windows multimedia timer is set to 1 ms as the minimum value, so that the resolution of measurement is to be 1 ms. Figure 13 shows the packet structure used in Bluetooth communication between the server PC and a mobile robot. Note that an execution time and 0 are given to T in control mode and in measurement mode, respectively. The size of a packet is fixed to 30 bytes. When the command packet is transmitted from the server PC to a mobile robot through the route (1) in Fig. 5, data at the offset positions from 0 to 11 are meaning and are newly written. On the other hand, when returning to the server PC from the mobile robot through the route (4), six PSD sensors' measurements are set to the offset positions from 12 to 29. It should be noted that the execution time in Fig. 9 is set to 0 ms in this measurement process.

Figure 14 shows a measurement result of dynamic sampling period T_k by using the proposed method shown in Fig. 12. It is observed that there exists some time dispersion around about 60 ms. Two causes can be considered. One is the delay of the Server PC side that occurs from the communication processing like Windows dialogue application → USB interface → Bluetooth device. The other is the delay of the mobile robot side that appears because of the A/D conversion of six PSD sensors and the internal communication processing like MicroConverter → Bluetooth device with 9600 bps. Further, when a DC motor built in each wheel is driven, undesirable spikes tend to appear according to the switching timing of ON/OFF. Consequently, the spikes bring out communication errors between MicroConverter and Bluetooth device on the mobile robot.

The value of about 60 ms is the pure round-trip delay time of this experiment setup, where Bluetooth and asynchronous serial communication of 9600 bps are used. It should be noted that the value itself of the time does not have critical meaning and worth because the dynamic sampling period depends on communication media and transmission

Offset	Type	Contents
0	ASCII	"I"
1	ASCII	"D"
2 – 3	ASCII	Robot No.
4	ASCII	Reflex act No.
5 – 11	ASCII	Exec. time (ms)
12 – 14	ASCII	PSD sensor 1
15 – 17	ASCII	PSD sensor 2
18 – 20	ASCII	PSD sensor 3
21 – 23	ASCII	PSD sensor 4
24 – 26	ASCII	PSD sensor 5
27 – 29	ASCII	PSD sensor 6

Fig. 13. Packet structure used in communication, in which a short execution time and 0 are given to "Exec. time (ms)" in control mode and in measurement mode, respectively

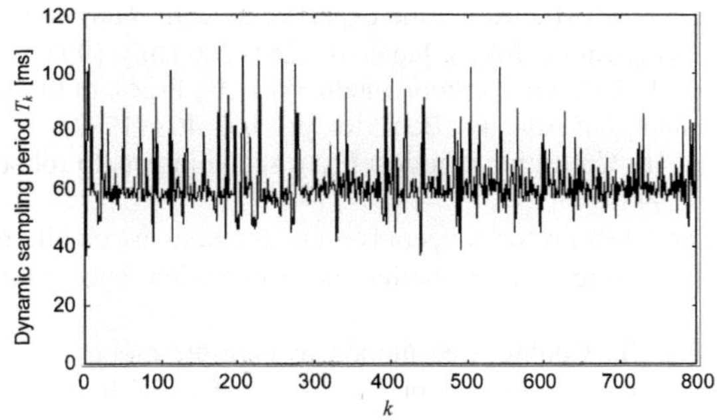

Fig. 14. An example of measurement result of dynamic sampling period T_k in case of the proposed multiple mobile robots system

line actually used. However, it is important to easily know the actual time required for one round-trip between a controller and a robot.

5 Conclusions

In this paper, a network-based subsumption architecture has been proposed not only to broadcastly control multiple mobile robots but also to reduce the total design cost for constructing the system. The multi-robots monitor running on the supervisory server and the agent dispatcher determining to which agent the highest priority of execution should be given, were described. When constructing an actual experimental system by using multiple mobile robots, there exits a serious demand to suppress the total design cost that is proportional with the increase of the number of robots. Even though each robot has only simple hardware platform with limitations of software development environment, the proposed system enabled to construct a high-level software architecture such as a schooling behavior. In addition, this system enabled the multiple mobile robots to be removed the complicated download process of hex code program in debugging work because the debugging was conducted only on a supervisory server, so that the high maintainability could be achieved. The basic processing capability of the network-based subsumption architecture should be experimentally evaluated in association with the measurement of the communication time. Because the communication time called the dynamic sampling period is an important criterion to know the throughput and/or the overhead. Therefore, a simple and reliable measurement method using a Windows multimedia timer and packet data was introduced. It is expected that the proposed multiple mobile robots system will be able to be used for a basic experimental system for swarm intelligence and its related fields.

References

1. Parsons, D., Canny, J.: A motion planner for multiple mobile robots. In: Procs. of IEEE International Conference on Robotics and Automation, Cincinnati, OH, USA, pp. 8–13 (May 1990)
2. Barman, A.R., Kingdon, S.J., Little, J.J., Mackworth, A.K., Pai, D.K., Sahota, M.K., Wilkinson, H., Zhang, Y.: DYNAMO: Real-time experiments with multiple mobile robots. In: Intelligent Vehicles Symposium, Tokyo, Japan, pp. 261–266 (July 1993)
3. Kube, C.R., Zhang, H.: Collective robotic intelligence. In: Procs. of the Second International Workshop on Simulation of Adaptive Behavior, pp. 460–468 (1992)
4. Kube, C.R., Zhang, H.: Collective robotics: From social insects to robots. Adaptive Behavior 2(2), 189–219 (1993)
5. Noreils, F.R.: An architecture for cooperative and autonomous mobile robots. In: Procs. of IEEE International Conference on Robotics and Automation, Nice, France, pp. 2703–2710 (May 1992)
6. Azarm, K., Schmidt, G.: Conflict-free motion of multiple mobile robots based on decentralized motion planning and negotiation. In: Procs. of IEEE International Conference on Robotics and Automation, Albuquerque, NM, USA, pp. 3526–3533 (April 1997)
7. Bennewitz, M., Burgard, W.: A probabilistic method for planning collision-free trajectories of multiple mobile robots. In: The Workshop Service Robotics - Applications and Safety Issues in an Emerging Market at the 14th European Conference on Artificial Intelligence, Berlin, Germany, 7 pages (August 2000)
8. Guo, Y., Parker, L.E.: A distributed and optimal motion planning approach for multiple mobile robots. In: IEEE International Conference on Robotics and Automation, Washington, DC, USA, pp. 2612–2619 (2002)
9. Parker, L.E.: The effect of heterogeneity in teams of 100+ mobile robots. In: Multi-Robot Systems, From Swarms to Intelligent Automata, vol. II, pp. 205–215. Kluwer Academic Publishers (2003)
10. Pimentel, B., Campos, M.: Cooperative communication in ad hoc networked mobile robots. In: IEEE/RSJ International Conference on Intelligent Robots and Systems, Las Vegas, NE, pp. 2876–2881 (October 2003)
11. Antonelli, G., Arrichiello, F., Chakraborti, S., Chiaverini, S.: Experiences of formation control of multi-robot systems with the null-space-based behavioral control. In: IEEE International Conference on Robotics and Automation, Roma, pp. 1068–1073 (April 2007)
12. Antonelli, G., Arrichiello, F., Chiaverini, S.: Experiments of formation control with multi-robot systems using the null-space-based behavioral control. IEEE Transactions on Control Systems Technology 17(5), 1173–1182 (2009)
13. Mondada, F., Bonani, M., Raemy, X., Pugh, J., Cianci, C., Klaptocz, A., Magnenat, S., Zufferey, J.C., Floreano, D., Martinoli, A.: The e-puck, a robot designed for education in engineering. In: 9th Conference on Autonomous Robot Systems and Competitions, Castelo Branco, Portugal, pp. 59–65 (May 2009)

14. Zhan, J.W., Cheng, W.Z., Chang, N.W.: Network-based multimodal human-robot interactions in ubiquitous computing environment. In: IEEE International Conference on Robotics and Biomimetics, pp. 131–136 (2009)
15. Han, X., Payandeh, S.: Experimental design and analysis in kinematic-based localization in wireless mobile platform network. In: 2012 IEEE International Systems Conference on Communication, Networking & Broadcasting; Components, Circuits, Devices & Systems; Computing & Processing (Hardware/Software), pp. 1–6 (2012)
16. Brooks, R.A.: A robust layered control system for a mobile robot. IEEE Journal of Robotics and Automation 2(1), 14–23 (1986)
17. Nagata, F., Yamashiro, T., Watanabe, K.: Cooperative swarm control for multiple mobile robots using only information from PSD sensors. Artificial Life and Robotics 16(1), 116–120 (2011)
18. Tosadenshi Ltd., `http://www.robot.tosadenshi.co.jp/`
19. Asakusagiken Co., Ltd., `http://www.robotsfx.com/main/list_controller.htm`
20. Nagata, F., Otsuka, A., Watanabe, K.: Network-based subsumption architecture for multiple mobile robots system. In: Proceedings of the 6th International Conference on Soft Computing and Intelligent Systems and the 13th International Symposium on Advanced Intelligent Systems (SCIS-ISIS 2012), pp. 187–192 (2012)
21. Watanabe, K., Shiraishi, Y., Tzafestas, S.G., Tang, J., Fukuda, T.: Feedback control of an omni-directional autonomous platform for mobile service robots. Journal of Intelligent & Robotic Systems 22(3/4), 315–330 (1998)
22. Nagata, F., Otsuka, A., Watanabe, K., Habib, M.K.: Multiple mobile robots system with network-based subsumption architecture. International Journal of Mechatronics and Manufacturing Systems 6(1), 57–71 (2013)
23. Otsuka, A., Nagata, F., Okino, T.: Measurement of dynamic sampling period of multi mobile robots system with wireless communication. In: Proceedings of 2013 IEEE International Conference on Mechatronics and Automation (ICMA 2013), pp. 1141–1146 (2013)

Expressions of Emotions of Koala Robot Based on Laban Movement Analysis

Fuminori Souma[1], Hiroyuki Masuta[2], and Hun-ok Lim[2]

[1] Graduate School of Engineering, Kanagawa University, Yokohama, Japan
r201170037tz@kanagawa-u.ac.jp
[2] Department of Mechanical Engineering, Kanagawa University, Yokohama, Japan
holim@kanagawa-u.ac.jp

Abstract. This paper describes the mechanism and the control system of a koala robot which is able to express various kinds of emotions based on the Laban Movement Analysis. The koala robot consists of a head, a body, and four legs. Its height is 370[mm], and its weight is 1.68[kg]. In order to react to an action, the robot should have a lot of sensors such as a force sensor, a distance sensor, and an infrared sensor. Various emotional expression experiments are conducted and evaluated based on questionnaires made by nineteen subjects.

Keywords: Mechanism, Pet robot, Emotion expression, Laban movement.

1 Introduction

Recently, robots that can communicate with human beings, such as pet robots, have been increasing in popularity due to the development of cutting edge technologies. These robots are expected to help people such as the elderly and those with disabilities. In order for humans to coexist with amusement robots, they must provide humans with a sense of affinity, relief and familiarity. Moreover, they should be able to verbally communicate with humans.

A baby seal pet-therapy robot (Paro), has several sensors that allow it to communicate with human beings [1]. Its length is 570 [mm], and its weight is 2.7 [kg]. The effect of the robot on humans was investigated in medical and welfare facilities. The results showed that it has a high therapeutic effect. A robotic dog (AIBO) that is capable of barking like a real pet was developed [2]. The robot can mimic various actions and emotions. Its length is 319 [mm], and its weight is 1.6 [kg]. The selling of AIBO terminated in 2006. NeCoRo that is a robotic cat was developed [3]. Its length is 320 [mm], and its weight is 1.6 [kg]. Its emotions change based on its interaction with humans. Furby that is a robotic owl was developed [4]. It is covered with fur and has a light sensor, an infrared sensor, and a microscope. Nabaztag that can provide us with information such as weather forecasts and traffic updates was proposed [5]. PLEO that is a robotic dinosaur was developed [6]. It is a commercially available pet robot. Its height is 300 [cm], and its weight is 1.5 [kg].

Y.-T. Kim, I. Kobayashi, and E. Kim (eds.), *Soft Computing in Advanced Robotics*, Advances in Intelligent Systems and Computing 269,
DOI: 10.1007/978-3-319-05573-2_2, © Springer International Publishing Switzerland 2014

Robots that can convey emotions through facial expressions have also been developed. ifBOT can create facial expressions by using LEDs [7]. MEXI is able to predict the emotions of human beings by using the fuzzy theory, and is able to express its own emotions by moving different parts of its face [8]. SAYA can express various emotions through facial expressions and neck movements [9, 10]. HRP-4C can express emotion using hands and feet movements and facial expressions [11]. KOBIAN can express seven different kinds of emotions using combinations of the face, legs, and arms [12]. However, these robots are not enough to give familiarity, affinity and relief to a human being. So, we have developed a koala-like robot. It has a force sensor, a distance sensor and an infrared sensor. It is able to express various kinds of emotions based on the Laban Movement Analysis.

2 Mechanisms of Koala Robot

2.1 Specifications of the Robot

Fig. 1 shows the whole view of the koala robot. Table 1 shows the specifications of the koala robot. We specifically chose a koala because of its cute appearance and friendly personality, which will give humans a sense of relief. The koala robot is composed of a head, a body, and four legs. Its height is 350 [mm], its width is 245[mm], and its length is 240 [mm]. The body is made out of aluminum alloy, acrylic resin, and acrylonitrile butadiene styrene. The acrylic hemisphere frame is used for its head to retain the koala's appearance. The koala robot wears a koala suit. The sensors, actuators, and battery are placed inside the robot.

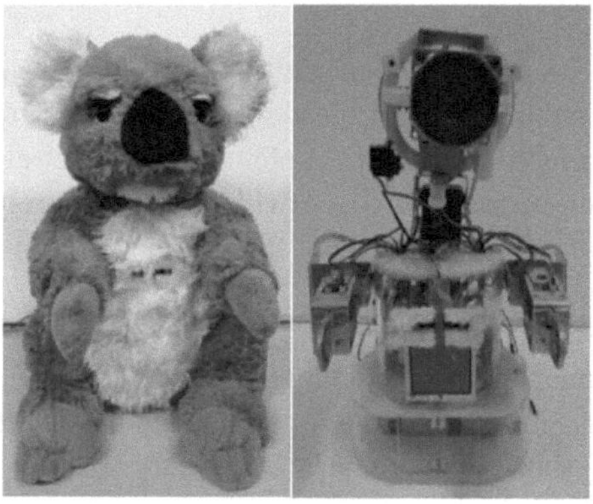

Fig. 1. Koala Robot

Table 1. Specifications of koala robot

Length	350[mm]
Width	245[mm]
Depth	240[mm]
Actuator	DC servo motor×8
Total Mass	1.68[kg]

2.2 Sensors

Fig.2 shows the layout of sensors and modules installed in the koala robot. The force sensors are installed in the head, shoulders, hands, abdomen and back. When a user strokes the head of the koala robot, the robot will express happiness. Moreover, when the user presses the robot's head or abdomen, the koala robot will express anger. In addition, the robot will offer to shake hands or wave when the user touches the shoulders or hands of the robot. The distance sensor is installed in the abdomen. The koala robot holds an object using distance information. The infrared (IR) sensor is installed at the tip of the jaw. The IR sensor is used to detect the proximity of humans. The acceleration sensor is installed in the body. The sound recognition system and the voice synthesis system are installed in the body. The loudspeaker is installed in the nose. Using the sound and voice system, the koala robot can communicate verbally with humans. When a user speaks to the koala robot, the robot will greet the user with a "good morning" or "hello."

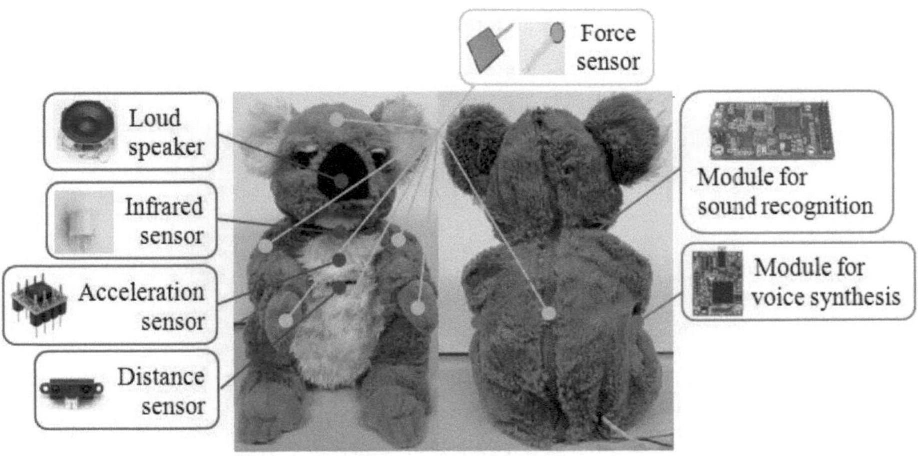

Fig. 2. Sensors and modules Installed in Robot

3 Emotion Expression

We apply the Laban Movement Analysis to express emotions. The Laban Theory consists of two components: "effort" and "shape." While the former describes the mechanical feature, the latter describes the geometrical feature. Furthermore, the two components are divided into two types of configurations: "fight" and "intoxication." "Fight" describes a powerful, spasmodic, straight line trajectory movement. "Intoxication" describes a weak, continuous, curved trajectory movement. In this research, we base the behavior of the robot on the "shape" configuration and develop four emotion expressions.

Fig. 3, 4, 5 and 6 show the snapshots of the emotion expressions. According to the Laban Movement Analysis, pleasure is expressed by spreading, raising and lowering the arms. Anger is expressed by folding and spreading the arms. Sorrow is expressed by lowering the head and arms and shrinking the arms. Enjoyment is expressed by raising each arm randomly.

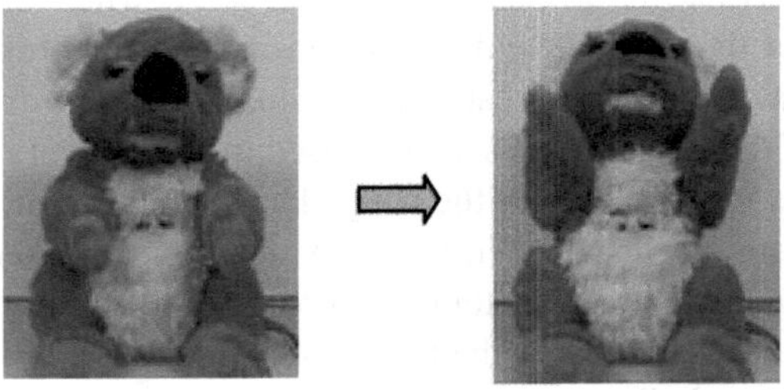

Raising and lowering the head and arms.

Fig. 3. Expressing Pleasure

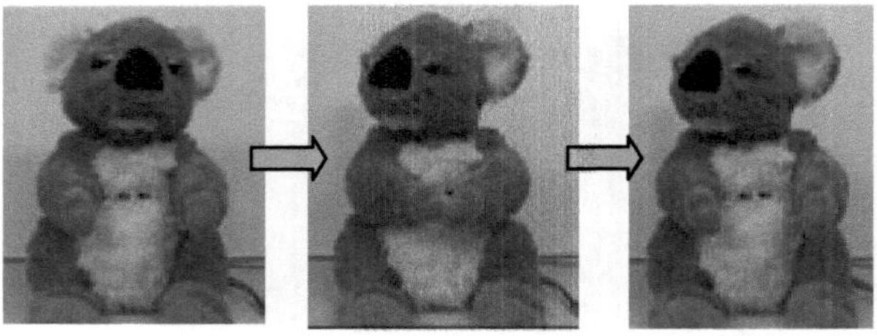

Folding the arms and spreading them vigorously, while shaking the neck.

Fig. 4. Expressing Anger

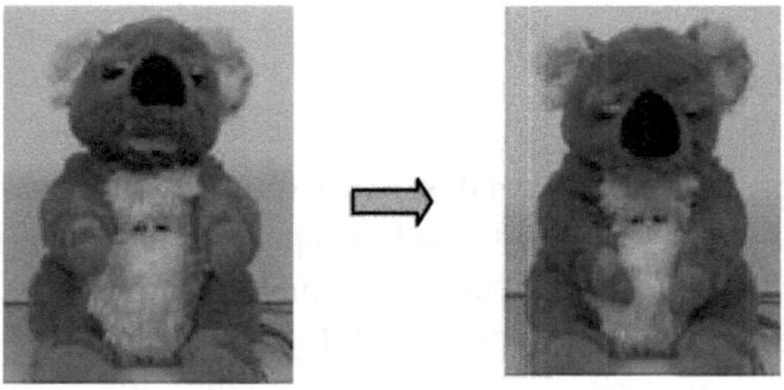

Lowering the head and arms and shrinking the arms.

Fig. 5. Expressing Sorrow

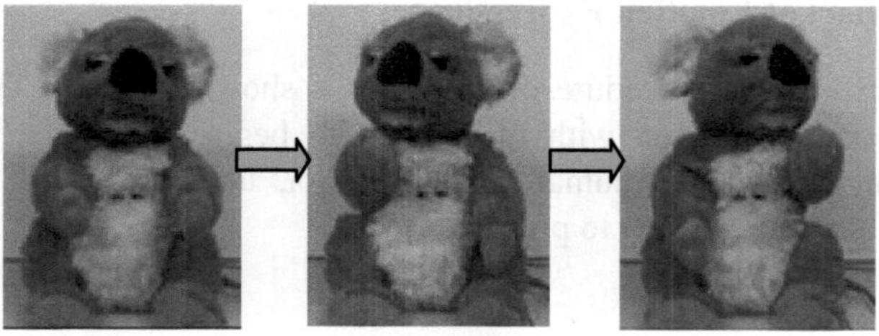

Raising each arm randomly.

Fig. 6. Expressing Enjoyment

4 Control of the Robot

4.1 Control System

A microcomputer board (Arduino MEGA 2560 R3) that is based on the ATMEGA1280 is used to control DC servo motors and voice synthesis devices. A microcomputer has 54 digital input/output pins, in which 6 can be used as PWM outputs, 16 analog input/output pins, a crystal oscillator which operates at 16 [MHz], a USB port, a power jack, an ICSP header, and a reset button. The speed of the DC servo motors is controlled by a Pulse Width Modulation (PWM). Also, a microcomputer board, SH7125F, is used to activate the module for sound recognition through serial communication. Fig. 7 shows the control system of the robot.

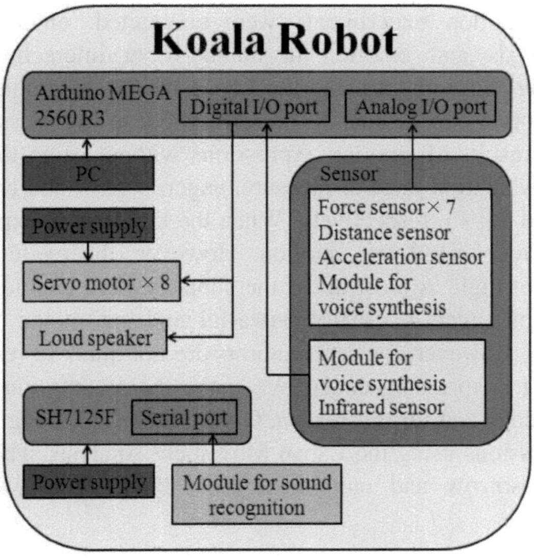

Fig. 7. Control System

4.2 Flow Chart of Emotion Expressions

The emotion expression procedure of the robot is shown in Fig.8. The koala robot waits for the communication with the human to begin. When the sensors detect communication between the human and the robot, the koala robot expresses the corresponding action according to prepared rules.

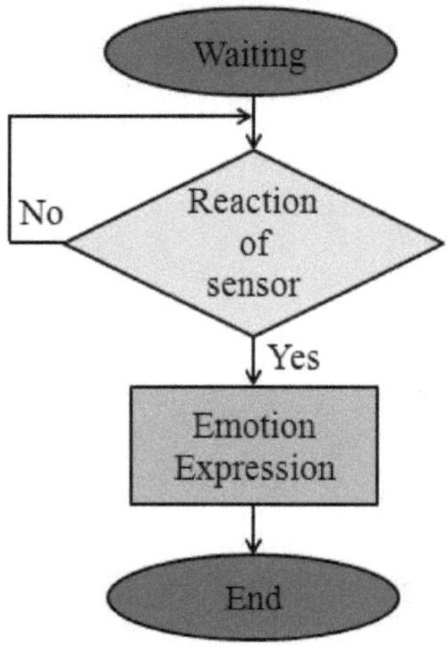

Fig. 8. Flow Chart of Emotion Expressions

5 Experiments and Results

Two kinds of emotion experiments were conducted: emotion with the Laban Movement Analysis and emotion in human-robot interaction. Firstly, emotion experiments using and without using the Laban theory were conducted. We showed the videos of emotion expressions of the koala robot to twenty subjects aged 10 to 70. Fig.9 shows the results of emotion expressions without using the Laban Movement Analysis. The evaluation rates of pleasure, anger, sorrow and enjoyment are 63[%], 23[%], 95[%] and 76[%], respectively. When the koala robot expressed sorrow, most of the subjects recognized the emotion. However, the evaluation rates of other emotions were not high. Almost all of the subjects were unable to recognize anger. The expression of anger is usually powerful and spasmodic with a straight line trajectory motion. However, the servo motors are not quick to respond and therefore, there is a delay in expressing anger. We make the assumption that the expression of anger is misrecognized for this reason. On the other hand, Fig.10 shows the results of emotion expressions using the Laban Movement Analysis. The evaluation rates of pleasure, anger, sorrow and enjoyment are 78[%], 77[%], 99[%] and 100[%], respectively.

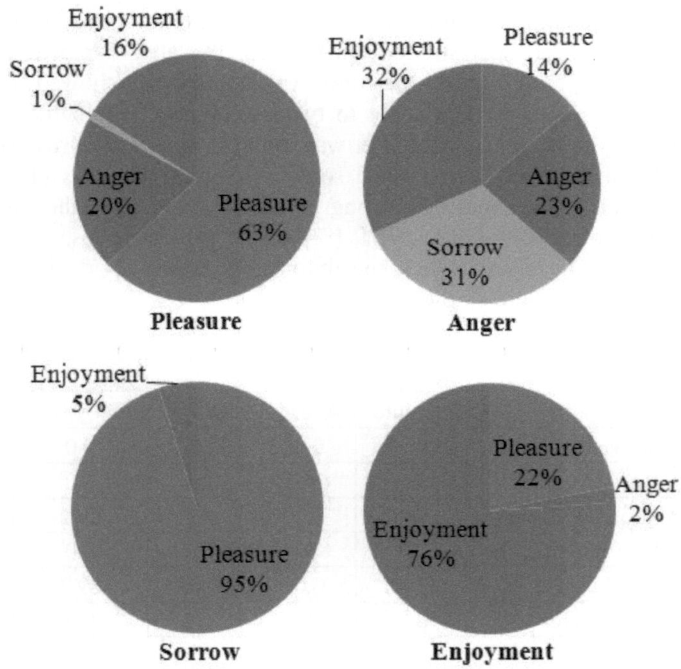

Fig. 9. The Results of Experiments without Laban Movement Analysis

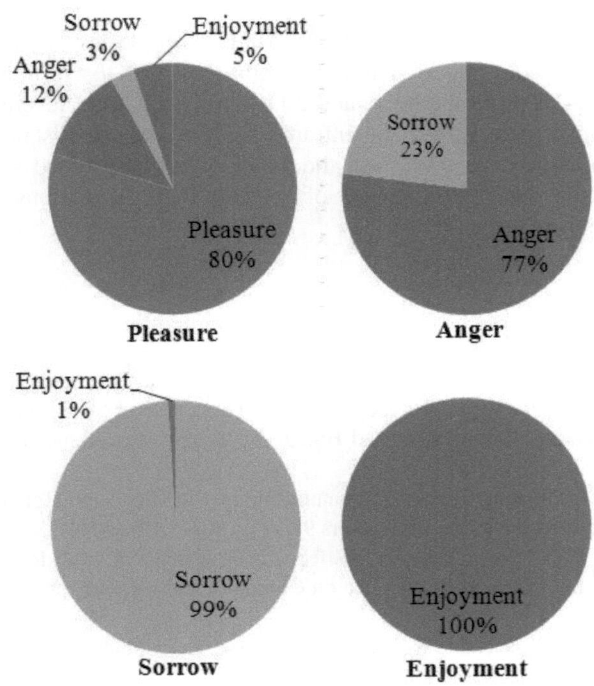

Fig. 10. The Results of Experiments with Laban Movement Analysis

Secondly, six kinds of experiments for human-robot relationship were conducted: emotions when the head was stroked and beaten, the abdomen was stroked and beaten, the back was touched, and the robot was slanted. We showed the videos of emotion expressions of the koala robot to twenty subjects. The subject selects the emotion that seems most appropriate after watching the videos. The results are shown in Table 2. When the head was stroked softly, all of the subjects chose pleasure. When the abdomen of the robot was strongly pressed, 84.6[%] of the subjects chose anger. When the robot was slanted, 76.9 [%] of the subjects chose sorrow. On the other hand, when the head was beaten, the abdomen is stroked, or the back is touched, highest evaluation rates were 69.2 [%].

Table 2. Evaluation of emotion expression

	Pleasure	Anger	Sorrow	Enjoyment
Stroke head	100.0	0.0	0.0	0.0
Beat head	0.0	69.2	30.8	0.0
Stroke abdomen	30.8	0.0	0.0	69.2
Beat abdomen	0.0	84.6	15.4	0.0
Touch back	30.8	0.0	0.0	69.2
Slant body	0.0	23.1	76.9	0.0

6 Experiments and Results

A koala robot that is able to express various kinds of emotions through verbal and nonverbal actions was developed. The koala robot has 8-DOFs in total, such as 2-DOFs in the face, 2-DOFs in the neck, and 2-DOFs in each arm. Several sensors and devices such as the force sensor, the infrared sensor, and the distance sensor are attached. The emotion expressions of the koala robot are based on the Laban Movement Analysis. The effectiveness of the emotion expressions was verified through emotion experiments.

References

1. Wada, K., Ikeda, Y., Inoue, K., Uehara, R.: Development and Preliminary Evaluation of a Caregiver's Manual for Robot Therapy using the Therapeutic Seal Robot Paro. In: IEEE International Symposium on Robot and Human Interactive Communication, pp. 533–538 (2010)
2. Fujita, M.: On Activating Human Communications with Pet-Type Robot AIBO. IEEE Institute of Electrical Electronics Engineers 92(11), 1804–1813 (2004)
3. Nakashima, T., Fukutome, G.: Healing Effects of Pet Robots at an Elderly-Care Facility. In: IEEE/ACIS International Conference on Computer and Information Science, pp. 407–412 (2010)
4. Breazeal, C.: Toward Sociable Robots. Robotics and Autonomous Systems 42, 167–175 (2003)

5. Tomitsch, M., Kappel, K., Lehner, A., Grechenig, T.: Towards A Taxonomy For Ambient Information Systems. In: Workshop on the Issues of Designing and Evaluating Ambient Information Systems (2007)
6. Jacobsson, M.: Play, Belief and Stories about Robots: A Case Study of a Pleo Blogging Community. In: IEEE Symp. Robot and Human Interactive Communication (RO-MAN 2009), pp. 232–237 (2009)
7. Matsui, Y., Kanoh, M., Kato, S., Nakamura, T., Itoh, H.: A Model for Generating Facial Expressions Using Virtual Emotion Based on Simple Recurrent Network. Journal of Advanced Computational Intelligence and Intelligent Informatics 14(5), 453–463 (2010)
8. Esau, N., Kleinjohann, L., Kleinjohann, B.: Integration of Emotional Reactions on Human Facial Expessions into the Robot Head MEXI. In: IEEE/RSJ International Conference on Intelligence Robots and Systems, pp. 534–541 (2007)
9. Hashimoto, T., Kobayashi, H.: Study on Natural Head Motion in Waiting State with Receptionist Robot SAYA that has Human-like Appearance. In: IEEE Workshop on Robotic Intelligence in Informationally Structured Space, pp. 93–98 (2009)
10. Hashimoto, T., Hiramatsu, S., Tuji, T., Kobayashi, H.: Realization and Evaluation of Realistic Nod with Receptionist Robot SAYA. In: IEEE International Conference on Robot and Human Interactive Communication, pp. 326–331 (2007)
11. Nakaoka, S., Kafita, S., Yokoi, K.: Intuitive and Flexible User Interface for Creating Whole Body Motions of Biped Humanoid Robots. In: IEEE/RSJ International conference on Intelligent Robots and Systems, pp. 1675–1682 (2010)
12. Zecca, M., Endo, N., Momoki, S., Itoh, K., Takanishi, A.: Design of the humanoid robot robot KOBIAN – preliminary analysis of facial and whole body emotion expression capabilities –. In: IEEE-RAS International Conference on Humanoid Robots, pp. 487–492 (2008)

Development of Steering Controller for Electric Vehicle

Daeyeong Im[1,*] , Hyunrok Cha[1], and Youngjae Ryoo[2]

[1] Automotive Components R&D Group, Korea Institute of Industrial Technology
{dylim,hrcha}@kitech,re.kr
[2] Department of Control Engineering and Robotics, Mokpo National University
yjryoo@mokpo.ac.kr

Abstract. In this paper, a steering system for a magnetic field based autonomous driving vehicle was designed and developed. A new velocity tracking frequency control method was applied for considering the velocity and torque characteristics of a stepping motor. For verifying the excellency of the velocity tracking frequency control method, this method was compared with a step control method.

Keywords: Steering mechanism, Unmanned driving, Unmanned vehicle.

1 Introduction

According to increases in the interest of autonomous vehicle driving throughout the world based lots of technologies have been developed.

A representative method in such driving is using camera vision and that has largely been studied [1,2]. In other technologies, some methods rely on using global positioning systems (GPS) and infra red radar sensing sensors have also been studied [3-5].

In America, a magnetic field-based autonomous driving system was first introduced using a PATH (California Partners for Advanced Transit and Highways) program in Berkely University sponsored by the California Department of Transportation (CALTRANS) in 1998 [6-8].

In the case of Netherlands, the companies of Frog Navigation System, which produces automated guided vehicles (AGV) for automation, and APTS, which is a sister company of the VDL group and produces large trucks and buses, developed magnetic marker-based public transportation buses and park shuttle buses in markets [9].

Then, magnetic field sensors are attached to measure the magnetic field generated from these magnets and control steering angles through measuring cross track errors in a vehicle for its longitudinal direction in order to develop an autonomous driving system.

In the advantages of the magnetic field-based autonomous driving system, the system is not affected by lights and weathers and accurately recognizes the present position of a vehicle in its driving.

In this study, a velocity tracking frequency control method is proposed to provide the proper velocity and torque for controlling steering angles. This steering system

* Corresponding author.

Y.-T. Kim, I. Kobayashi, and E. Kim (eds.), *Soft Computing in Advanced Robotics*,
Advances in Intelligent Systems and Computing 269,
DOI: 10.1007/978-3-319-05573-2_3, © Springer International Publishing Switzerland 2014

and velocity tracking mechanism developed in this study are applied to an autonomous driving vehicle in order to verify its practicability. This study begins in Section 2 by describing the design and development of a steering system. Section 3 applies a test for the developed steering system and analyzes the measured data. Section 4 installs the developed steering system to an autonomous driving vehicle and analyzes the measured data. Section 5 verifies the advantage and disadvantage of the system through examining the conclusion of this study.

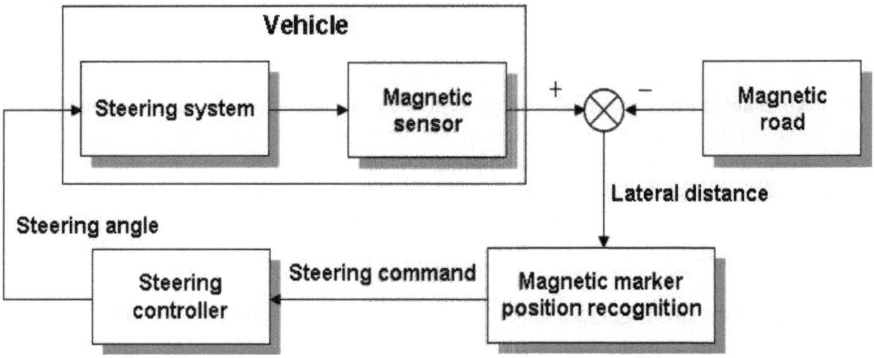

Fig. 1. Block diagram of unmanned vehicle system based on magnetic marker

2 Steering Actuator

2.1 Design of Steering Actuator

In this section, a steering control system is designed using a stepping motor. Regarding the advantages in a stepping motor, it is able to perform a precise angle control in a stepping according to frequencies and represents no accumulation of the error of rotation angles and large holding torques. Also, the stepping motor represents easy maintenances and uses because it has no contact points in motor brushes. Regarding the frequency characteristics in this motor, it shows an increase in rotation speeds according to the increase in frequencies but that decreases its torques. Also, the low frequencies exhibit low torques but that increases its torques.

2.2 Algorism of Steering Control

Shown in Fig. 3 the steering system consists of five sections. The sensor used in this system outputs the lateral position data that is obtained as the vehicle is broken away from a magnetic field lane. Regarding the lateral position data output from the sensor, the steering command data for implementing an actual steering angle can be calculated based on the reference value and error of the sensor under an assumption that the vehicle is always positioned at the center of the magnetic field lane.

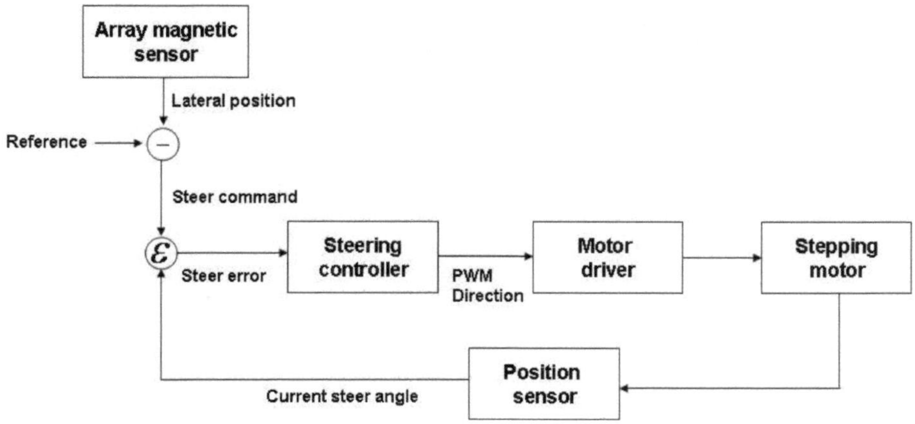

Fig. 2. Block diagram of steering system

2.3 Frequency Control of Stepping Motor

As the center of gravity in a vehicle is moved to the real during driving, the higher velocity of the vehicle would make it easier to control the steering angle.

However, as a vehicle is stopped or driving with a low velocity, the frequency is to be decreased in order to increase its steering torque because the center of gravity is moved to the front.

That is, the frequency of controlling the rotation speed and torque in the steering stepping motor is to be varied according to the condition of the vehicle.

In addition, it is to be varied according to the velocity of the vehicle and the number of passengers in the vehicle or the condition of roads.

Also, the steering control is smoothly performed during driving. As the vehicle is approached to the target value of the steering, tremors in the vehicle are to be reduced by preventing an overshoot through a slow approach to the target value.

Thus, in this study, a variable frequency control method for the frequency control of the stepping motor is proposed.

As shown in Fig. 3, the variable frequency control method does not implement the steering control because the current error (e_t) agrees with the magnetic field lane in a Dead Zone. The frequency is to be varied as $F_l \sim F_h$ in a range of $e_{min} \sim e_{max}$ and rapidly controlled to make it enters a stable region with maximum velocity as the error is more than e_{max}. In addition, the system is configured to operate it with a frequency of F_l as the vehicle is stopped or driving with a low velocity.

Fig. 4 represents a practical frequency control method. If the value of ε generated by the difference between the current position sensor of the vehicle and the measured data in the sensor is positioned at a range of 10~512, the frequency will be controlled at a range of 500~2000. Also, if the value is more than 512, the frequency will be fixed at 2000.

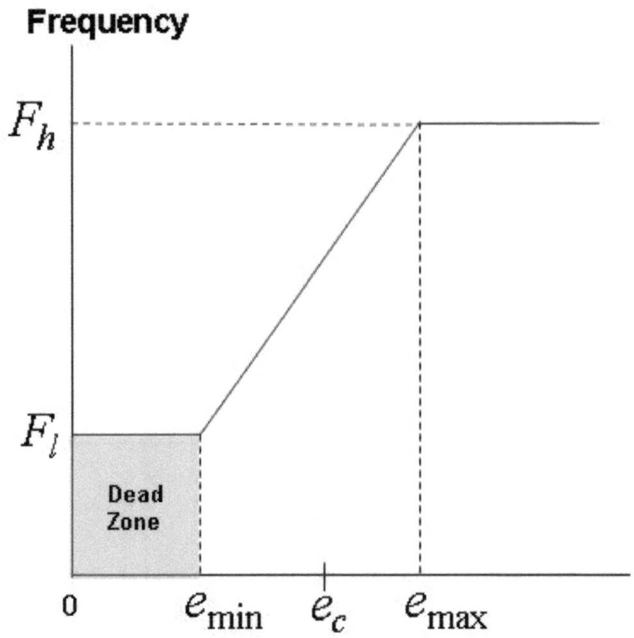

Fig. 3. Concept of frequency control for steppingping motor

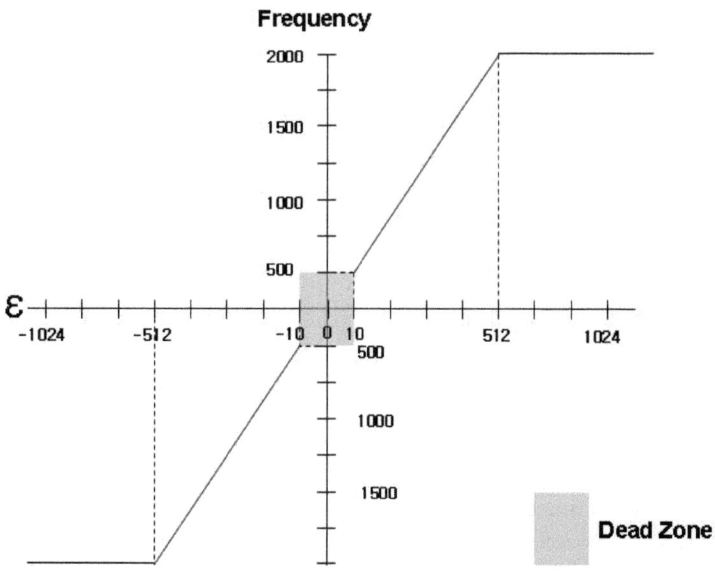

Fig. 4. Frequency control for steppingpingmotor

The reason to fix it 2000 is due to the fact that if the frequency is determined by more than 2000, it will cause a break away from the magnetic field lane due to the lack of torques for controlling the vehicle.

$$\mathrm{F} = \frac{e_c - e_{min}}{e_{max} - e_{min}} \times (F_h - F_l) \quad (1)$$

$$\mathrm{Frequency} = \mathrm{F} + F_l \quad (2)$$

3 Experimental

3.1 Steering Actuator

A steering system presented in Fig. 5 was developed to verify the practicability of the designed steering system. The exited On/Off signals calculated by the steering controller for the rotation speed and direction of the motor were transmitted to the stepping motor. he steering angle for left and right sides were 20[˜] and 20[˜], respectively. As the pulley of the motor shaft was rotated 5 turns, the pulley of the handle shaft was rotated 2.5 turns. A 2:1 timing pulley was installed at both the motor shaft and the handle shaft. The pulleys installed at the handle and motor shafts were connected using a high tension timing belt that has a strong resistivity for changes in temperature and external environments. For adjusting the tension of the belt, the motor can be moved to a horizontal direction while the handle shaft is fixed. A high precision position sensor, M-1035 by Copal Electronics, was installed at the center of the rear face of the motor in order to measure the current steering angle of the vehicle. A coupler was installed between the central axes of the motor and the position sensor in order to protect the sensor. he handle was developed for both automatic and manual modes.

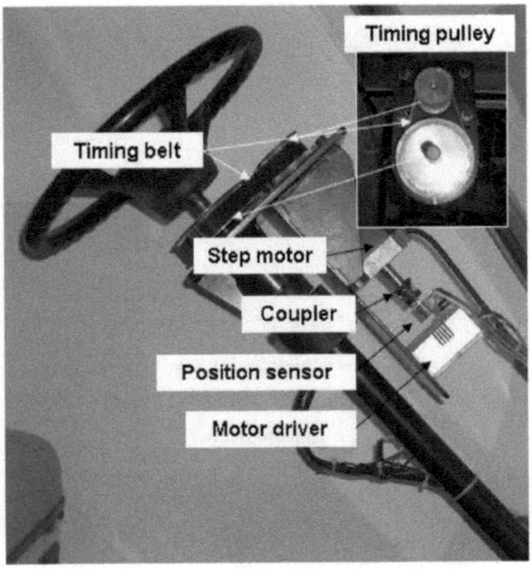

Fig. 5. Steering Actuator

3.2 Experimental of Steering Control

In this section, the stepping control method and the developed frequency control method are compared in a specific experimental environment for verifying the applicability of the steering system developed in this study. Regarding the environment, the sensor with a width of 90[cm] was installed at the bumper of the vehicle and set at a distance of 15[cm] from the ground. Magnetic markers were made of neodymium with a diameter of 2.5[cm] and a height of 2[cm]. Magnets were moved from the left end of the sensor to the right end and four different data, such as Steering command, Steering angle, Steer error, and Frequency, were measured according to the rotation of the steering system. The monitoring system used in this measurement was developed using the LabView.

3.3 Step Control

Fig. 6 represents the frequency control using a step control method. As the error was above 256, the frequency was about 2000[Hz] and that allows its rotation to the utmost speed. Here, the frequency per stepping was varied by 150[Hz] and the minimum frequency was determined by 650[Hz]. In the case of the stepping control, however, the steering control could not be smoothly performed because of varying the frequency with 150[Hz] and took a very long time, 6[sec], to approach to the target value while the current steering angle was tracking the steering command. In this case, the robot like vehicle cannot track the magnetic road and that may leave the lane.

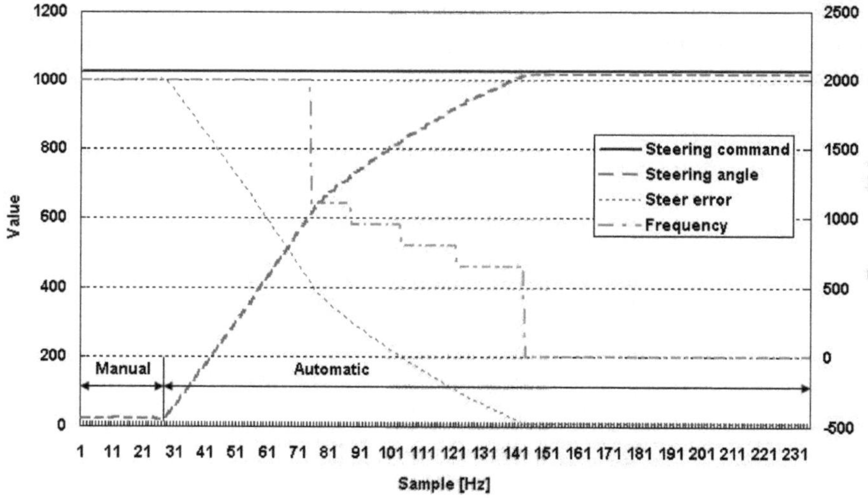

Fig. 6. Step Control

Fig. 7 shows the results of the velocity tracking frequency control proposed in this study. First, the frequency was controlled according to the measurement of the encoder pulse, which can be used to verify the velocity of the vehicle. That is, as the encoder pulse was below 2, the frequency was operated as 500[Hz]. As the encoder pulse was above 2, the frequency was linearly varied between 2000[Hz] maximum and 500[Hz] minimum according to the steering angle of 10~256 in which the smaller error of the steering angle would make it slower to move the system and the larger error would make it faster to move it. Thus, the error of the steering angle can be decreased as the vehicle is aligned to the magnetic field lane where the precision in the steering angle control can be improved as the control is carried out with a low velocity rather than a fast velocity. Based on the graph, the command of the steering angle exhibited no changes in its mode either the manual or the automatic mode with a constant value of 1024. The pulse and frequency were 0 and 500[Hz], respectively, because the vehicle was not started in the manual mode. Also, the steering angle and its error were 24 and 1000 respectively. The vehicle was started from the pulse of 44[sample], which begins the automatic mode, and the frequency was varied according to the error of the steering angle as the encoder pulse more than 2. Here, the operation was implemented up to the frequency of 2000[Hz] in order to present the fastest motion because the error was larger than the configured value of 256 and the frequency was varied from 2000 to 500[Hz] because the error was determined as 256~10 up to 104~137 samples. The steering angle reduces the error of the steering angle through tracking the command of the steering angle.

The number 10 in the error of the steering angle represented the configured Dead Zone. Thus, the steering angle tracked the command of the steering angle with the maximum velocity up to 43 ~ 103 samples and the time to approach to the 60[%] of the command was about 2.95[sec].

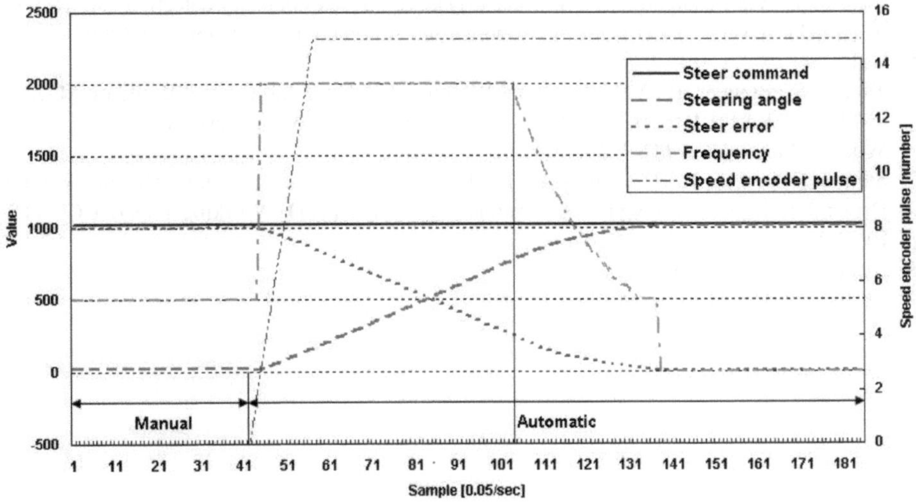

Fig. 7. Intelligent Control

In addition, it is necessary to increase the rotational force because the load applied to the system is large as the magnetic field lane is positioned at both ends of the sensor. In this case of the conventional method, it showed a derail. However, the proposed method represented no derails and allowed its control. Thus, it was verified that the velocity tracking frequency control method is a better way than the conventional way.

4 Conclusion

In this study, a steering system for a magnetic field based autonomous driving vehicle was designed and developed. A new velocity tracking frequency control method was applied for considering the velocity and torque characteristics of a stepping motor.

For verifying the excellency of the velocity tracking frequency control method, this method was compared with a stepping control method. It is necessary to apply an intelligent control method like neural networks or fuzzy control to the fields of vehicle control and vehicle dynamics in order to the maximum error within 5[cm] in future.

References

1. Desouza, G.N., Kak, A.C.: Vision for Mobile Robot Navigation: A Survey. IEEE Transactions on Pattern Analysis and Machine Intelligence 24(2) (2002)
2. Ryoo, Y.-J., Lim, Y.-C.: Visual Control of Autonomous Vehicle by Neural Networks Using Fuzzy-Supervised Leaning. Journal of Electrical Engineering and Information Science 2(2), 77–85 (1997)
3. Upadhyay, T., Cotterill, S., Deaton, A.W.: Autonomous GPS/INS navigation experiment for space transfer vehicle. IEEE Trans. on AES 29(3), 772–785 (1993)
4. Wijesoma, W.S., Kdagoda, K.R.S., Balasurjya, A.P.: Road-boundary detection and tracking using ladar sensing. IEEE Trans. on Robotics and Automation 20(3), 456–646 (2004)
5. http://www.path.berkeley.edu
6. Tan, H.-S., Bougler, B.: Vehicle Lateral Warning, Guidance and Control Based on Magnetic Marker: PATH Report of AHSRA Smart Cruise 21 Proving Tests, Calivornia PATH Working Paper, UCB-ITS-PWP-2001-6
7. Steinfeld, A., Tan, H.-S.: Preliminary Findings for a Lane-Keeping and Collision-Warning Driver Interface for Snowplow Operations, California PATH Working Paper, UCB-ITS-PWP-99-6
8. http://www.frog.nl
9. http://www.apts-phileas.com

Design of Cruise Control System for Electric Vehicle Using Piece-Wised Control

Yongjun Lee and Young-Jae Ryoo

Dept. of Control Engineering and Robotics, Mokpo National University
1666 Youngsan-ro, Cheongye-myun, Muan-Gun Jeonnam 534-729, Korea
{yongjun,yjryoo}@mokpo.ac.kr

Abstract. In this paper, a design scheme of a speed controller for an electric vehicle's cruise driving using a piece-wised control is proposed. Cruise driving of electric vehicle is one of the major performance functions. For the cruise driving, while ideal driving motors have linear characteristics of voltage to speed, practical motors have nonlinear characteristics of voltage to speed. The performance of nonlinearity can be improved through the closed-loop control to compensate it. In this paper, we have improved the performance by applying piece-wised control because the drive motor that has highly nonlinear characteristics is difficult to obtain adequate performance only using closed-loop control. In order to test the proposed method, the experiments were carried out by applying the proposed method after building the electric vehicle equipped with the drive motors that has highly nonlinear characteristics. The experiment results using the proposed piece-wised control show superior performance than that using closed-loop control.

Keywords: Cruise control, electric vehicle, piece-wised control.

1 Introduction

Lately, due to the drastic advance on the technology of the unmanned driving, there are increasing interests in the development of the manned and unmanned electric vehicles using electric power. Except the technological advances, manned and unmanned electric vehicle research and development is underway in related industries because rising oil prices and energy depletion, pollution problem is being aggravated. Especially, the energy issue of the fossil fuel is causing the economic problem which is serious at whole world and more and more invention of eco-friendly electric vehicles is important, because of energy policies of the United Nations Framework Convention on Climate Change. [1-4]

Unlikely usual vehicle that is running by oil energy, electric vehicle is running the motor to get power by using battery, and consist of rechargeable battery, driving motor, motor controller, gear, etc. By using a motor instead of the engine, the electric vehicle is less noisy, easy maintenance, eco-friendly but inferior driving performance, short mileage. To solve that disadvantage, the way that increases energy efficiency by using cruise control of electric vehicle needs. [5-11]

The electric vehicle is possible for driving by using the driving motor. For the design of the controller, we can do speed control to electric vehicles by calculating

Y.-T. Kim, I. Kobayashi, and E. Kim (eds.), *Soft Computing in Advanced Robotics*, Advances in Intelligent Systems and Computing 269,
DOI: 10.1007/978-3-319-05573-2_4, © Springer International Publishing Switzerland 2014

characteristic curve of the driving motor, applying linear characteristic in controllable speed range to drive motor controller. But the practical motor has non-linear operating characteristics, and occurs precise control error when the non-linear section operates at the performance characteristic of the driving motor.

Thus, we propose the cruise control system by using piece-wised control. The piece-wised control is the method that controls at section appeared linear characteristic by analyzing operating characteristic curve of the driving motor. To experiment proposed method, we consist of the vehicle that is equip with the motor that has high non-linear characteristic.

2 Development of Electric Vehicle

2.1 Structure of Experimental Electric Vehicle

For the experiment with the proposed method, the driving part was developed. The driving part of the experimental electric vehicle used the transaxle set that consist of the dc motor, differential gears, electric brake. The motor is 24V 320W, and it has a high non-linear characteristics. The electric brake begins to work when motor stop working, which make the vehicle drive safely.

2.2 Configuration of the Driving Part

For the driving of the vehicle, we deigned the drive controller through the microprocessor. The driving motor for the driving use the DC motor, and it was controlled by the DC motor driver.

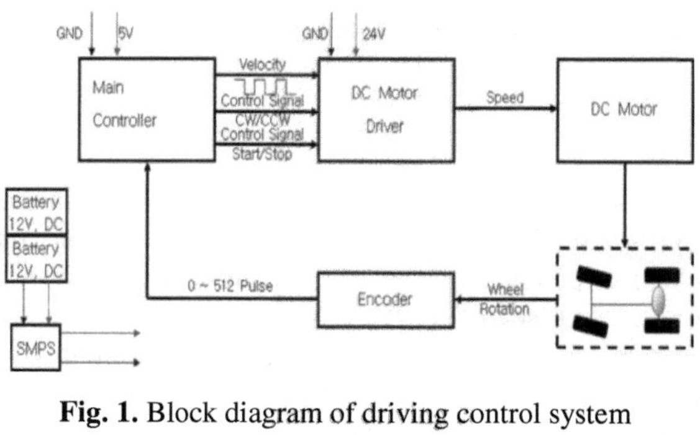

Fig. 1. Block diagram of driving control system

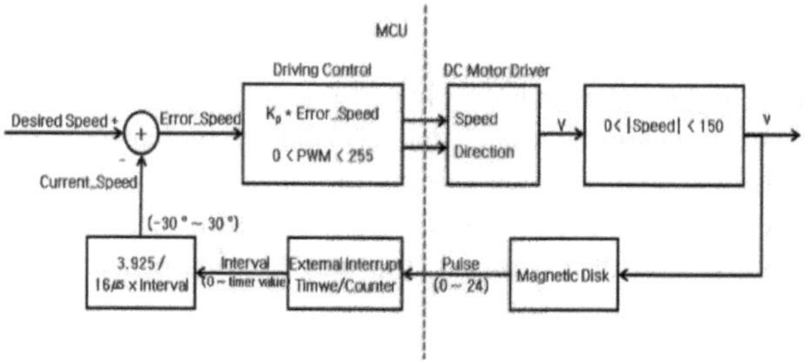

Fig. 2. Closed-loop driving control block diagram

The signal that occurred at the drive controller is divided into the pwm signal to determine the drive speed of the vehicle, drive START/STOP signals, and CW/CCW signal that determine the drive direction. after that, these are inputted to the DC motor driver. To measure the drive speed of the electric vehicle, we design and develop the magnetic disk, and measure the driving speed by attaching the magnetic disk to the wheel of the driving part.

2.3 Driving Speed Control System

In case of the developed electric vehicle in this paper, we have difficult installing the encoder because the space of the driving part was narrow. To overcome this, a magnetic disk is designed and developed as shown in Figure 3.

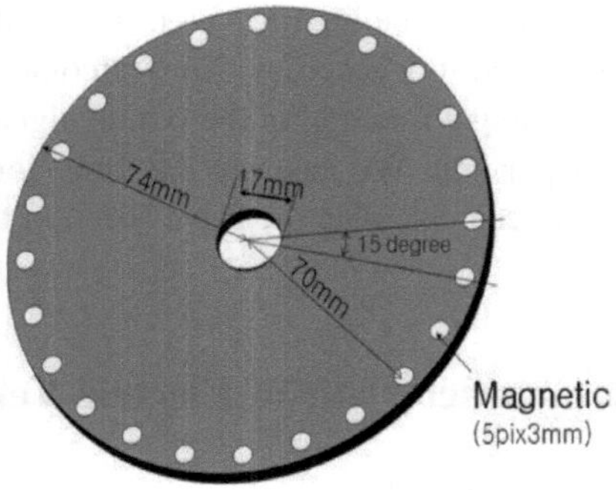

Fig. 3. Magnetic disk

The magnetic disk on the inside frame of the drive wheel is used for the measurement of the speed. It was produced in thin plate, therefore it can be installed easily in the confined space. According to the size of the wheel that we want to equip with, we can adjust the size of the disk and can decide the number of the attached magnet. We attach the magnetic disk to the wheel and the hall sensor to the body of the vehicle, and the position of the hall sensor is that the magnet should pass in the vertical direction next above the hall sensor.

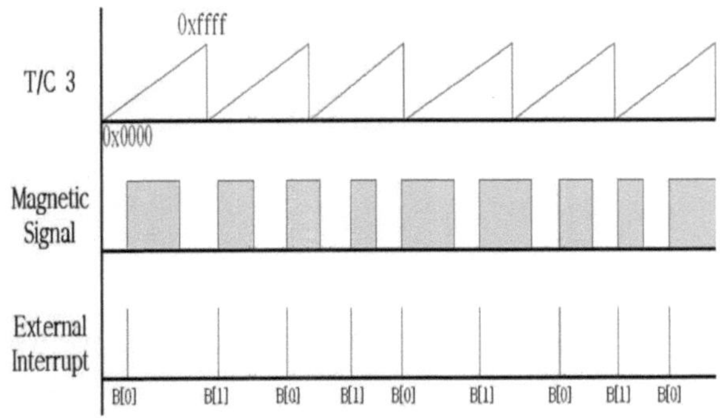

Fig. 4. Speed sensing using magnetic disk

The magnetic disk on the inside frame of the drive wheel is used for the measurement of the speed. According to the size of the wheel that we want to equip with, we can adjust the size of the disk and can decide the number of the attached magnet. It was produced in thin plate, therefore it can be installed easily in the confined space. We attach the magnetic disk to the wheel and the hall sensor to the body of the robot, and the position of the hall sensor is that the magnet should pass in the vertical direction next above the hall sensor.

According to the wheel's rotation, it is rotating together with the magnetic disk, and generate the signal during it pass over the hall sensor each time. At this time, the occurred signal was entered to the main controller. By using the signal entered, the main controller can measure the driving speed of the vehicle.

When we used the magnetic disk, the number of the pulses was less than using the encoder. So it has a problem to use the algorithms, it used calculus using the number of pulses per unit, because it had low accuracy. To overcome this, we use the way to measure the drive speed by using a time lag of the output signal occurrence instead of using the number of pulses per unit. We measure the time from the first signal to the next signal, and that time, we can calculate the drive speed of the vehicle by using the moving distance.

3 Driving Speed Characteristic of Electric Vehicle

3.1 Characteristic of Driving Motor

For the controller design of electric vehicle driving, Experimental operation of a DC motor was operated. In order to look for operating characteristic curve values of the DC motor, increasing PWM from 0 to 10 units motor was operated and the rotational speed of the driving wheel was measured. By analyzing the measured value of the speed, motor characteristic curve was calculated. Minimum PWM value for the DC motor to operate and the value of the proportional constant for speed control(kp) was set.

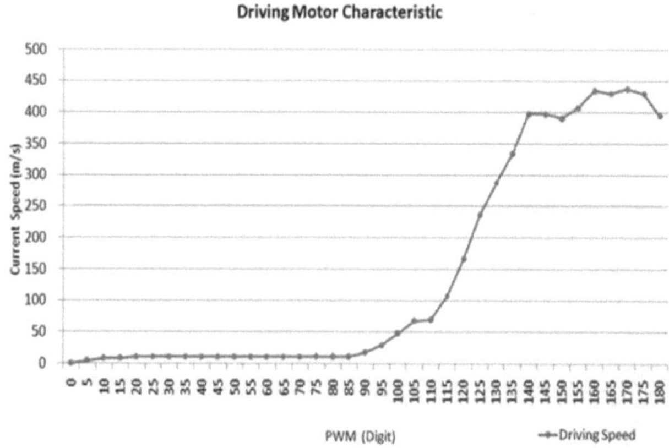

Fig. 5. Characteristic of DC motor

3.2 Characteristic of Open-Loop Control

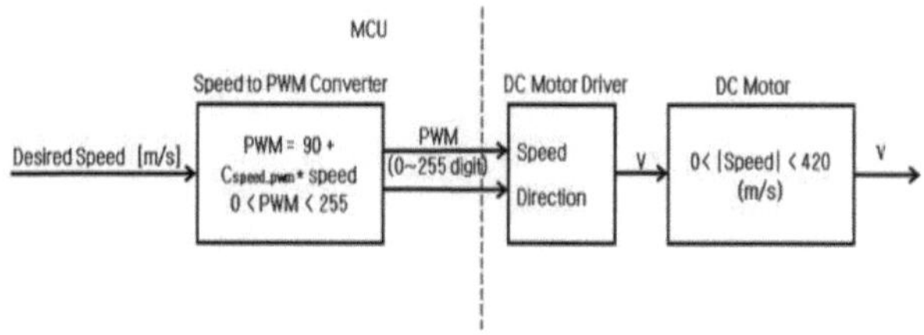

Fig. 6. Open-loop driving control block diagram

For the driving speed control, open-loop control experiment was conducted. We increase the value of the desired velocity and compare the value of the driving speed. Figure 7 is the result of the open-loop control experiment.

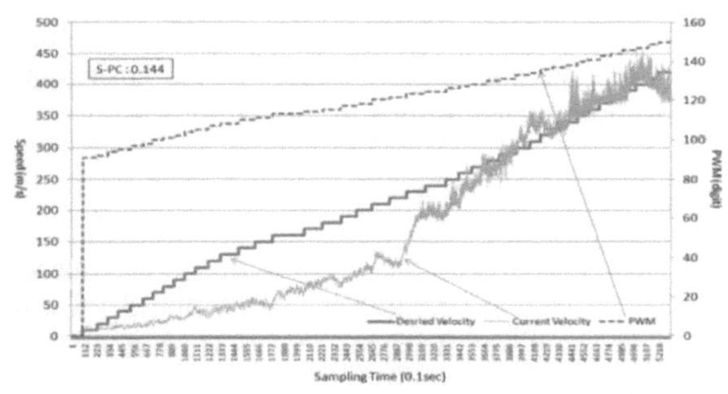

Fig. 7. Result of open-loop control

4 Controller Design and Experiment

To correct the error identified in the open-loop control experiment, we conduct the closed-loop control experiment that feedback the current driving speed value and compensate to the input.

4.1 Closed-Loop Proportional Control

We increase the value of the desired velocity gradually on a 10 unit basis, and compare with the measured driving speed value, and express as a graph. Figure 8 is a

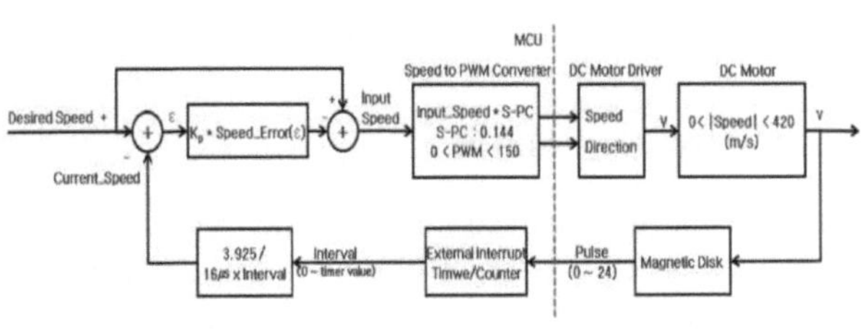

Fig. 8. Closed-loop driving control block chart

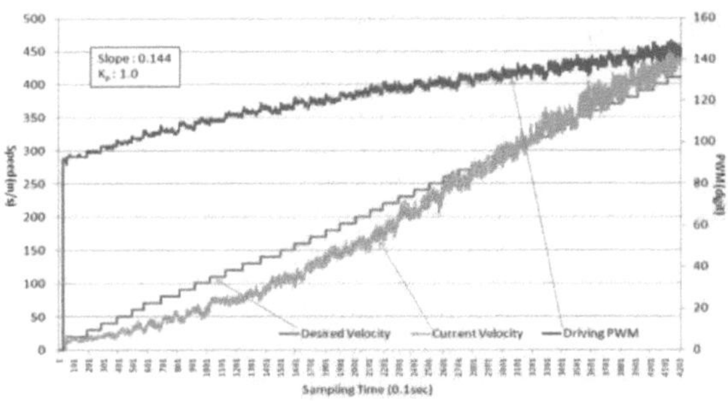

Fig. 9. Result of closed-loop control

block diagram of the Close-Loop control program. By using the magnetic disk in the input driving speed, the error of the current driving speed add, and generate the PWM using the S-PC.

Closed-loop control method can confirm the tracking closer to the target speed than open-loop control method. We corrected by feedback the error, but confirm that the error cannot reach to the input driving speed.

4.2 Design of Piece-Wised Controller

In the closed-loop control experiment, we control current driving speed values more closely to the target by revising the error. But, due to the S-PC value characteristic that was used to convert input speed values to PWM, it does not reach the target completely, and the error has occurred.

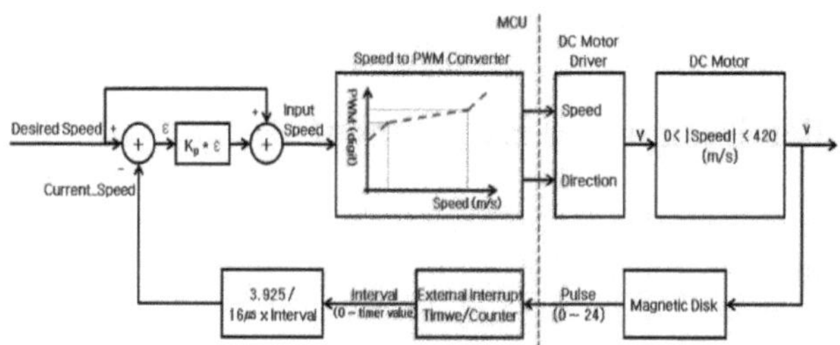

Fig. 10. Piece-wised control block chart

In order to revise these, we divided linear characteristic region in the motor characteristic region. Using the piece-wised control that depending on the size of the input speed the value of S-PC is applied differently, the controller was designed.

4.3 Piece-Wised Control Experiment

Figure 3 is the result of the piece-wised control experiment. We confirmed that the speed tracking is progressed to the target closely when compared the result of the open-loop control experiment with the closed-loop control experiment.

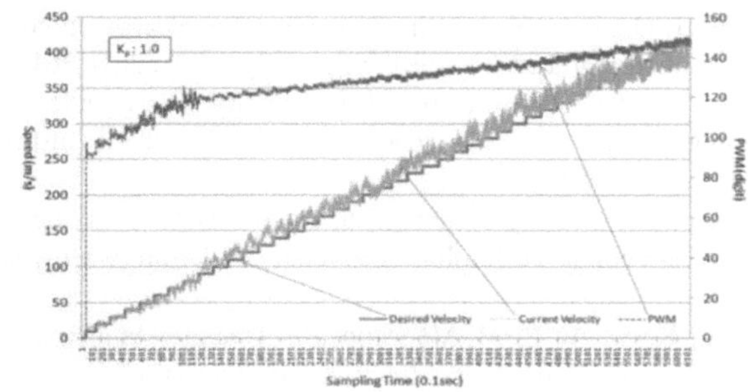

Fig. 11. Result of piece-wised control

5 Conclusion

In this paper, the design of the cruise control system of the electric vehicle by using piece-wised control system is proposed. For the experiment of the proposed speed controller, the electric vehicle is developed. Based on the characteristic experiment and analysis of the driving motor in the diving part, the speed control system is designed. After design of the piece-wised controller based on the experiment result of the operating characteristic of the driving motor, we conduct the experiment. The controller using the piece-wised control is confirmed to follow speed more closely to the target.

Acknowledgements. This research was financially supported by the Ministry of Education, Science Technology (MEST) and National Research Foundation of Korea(NRF) through the Human Resource Training Project for Regional Innovation (No. 2012026068)

References

1. Kim, G.-M., Ryoo, Y.-J.: Development of Automatic Guided Mobile Robot Using Magnetic Position Meter. In: Proceedings of World Academy of Science, Engineering and Technology, vol. 25 (2007)
2. Im, D.-Y., Kim, D.-Y., Lee, Y.-J., Ryoo, Y.-J., Cha, H.-R.: Unmanned Driving of Intelligent Robotic Vehicle. In: Proceedings of International Symposium on Advanced Intelligent Systems (2009)
3. Choi, K.-H.: Low-speed electric vehicles for driving high-efficiency, high-density development of the controller module. Department of Electrical Engineering, Graduate School of Chonnam National University, worthy scholar thesis for a degree (2012)
4. Lim, D.: Robotic vehicle using guidance system based on magnetic marker. Department of Control System Engineering Graduate School of Mokpo National University, doctor thesis for a degree (2008)
5. Do., H.: Design of centrifugal belt pulley for continuously variable transmission(CVT) of electric vehicle. Department of Control and Instrumentation Engineering, Graduate School of Industry, Pukyoung National University, worthy scholar thesis for a degree (2012)

6. Ban, D.-H., Park, J.-O., Lim, Y.-D.: Speed control for electric motorcycle using fuzzy controller. Journal of Korean Institute of Intelligent Systems 22(3), 361–366 (2012)
7. Lee, S.-H., Rhee, S.-Y.: A controller based on velocity estimator for a wheeled inverted pendulum running on the inclined road. Journal of Korean Institute of Intelligent Systems 21(3), 283–289 (2011)
8. Kim, H.: A study on the steering system of an electric vehicle having independent power source. Institute of e-Vehicle Technology University of Ulsan, worthy scholar thesis for a degree (2010)
9. Kim, Y.-S., Hong, K.-S.: An IMM Algorithm for Tracking Maneuvering Vehicles in an Adaptive Cruise Control Environment. International Journal of Control, Automation, and Systems 2(3), 310–318 (2004)
10. Jin, G.-G., Lee, H.-S., Lee, Y.-H., Lee, Y.-I., Park, Y.W.: Fuzzy-Based Speed Estimation for Navigation of Unmanned Robots. International Journal of Control, Automation, and Systems 8(2), 385–391 (2010)
11. Van Nguyen, Q., Eum, H.-M., Lee, J., Hyun, C.-H.: Vision Sensor-Based Driving Algorithm for Indoor Automatic Guided Vehicles. International Journal of Fuzzy Logic and Intelligent Systems 13(2), 140–146 (2013)

Vibration Minimization of Tower Typed 2-Wheeled Mobile Robot Using Acceleration and Deceleration Velocity Profile Method

Gyu-Jin Jo and Young-Jae Ryoo*

Department of Control Engineering and Robotics, Mokpo National University
1666 Youngsan-ro, Cheonggye-myeon, Muan-gun, Jeollanam-do, 534-729, Korea
{jjojjo,yjryoo}@mokpo.ac.kr

Abstract. In this paper, a vibration minimizing technique of a tower typed 2-wheeled mobile robot using acceleration and deceleration velocity profile method is proposed. Even though very low vibration on the body of a 2-wheeled mobile robot, it affects to the serious vibration on the top of the tower. We proposed the velocity profile method with acceleration and deceleration to minimize the vibration of the 2-wheeled mobile robot. The proposed method is experimented with three axis gyro sensors and three axis acceleration sensors attached on the robot's top to measure the vibration and analyze the effects of velocity profile.

Keywords: tower type, 2-wheeled mobile robot, velocity, profile method.

1 Introduction

Recently, tower type two-wheeled mobile robots were developed for guidance or docent robots. [1] The lower part of these robot's structure has wheels and the upper part has display devices. If the vibration occurs at the mobile robot, it is hard to see the display on the top of the robot.

Therefore, we need the method to decrease the vibration. As a previous research to reduce vibration, methods to remove the robot's vibration by using the mechanism were proposed. [2, 3] A technique to decrease vibration by using the vibration controller disturbance observer was proposed [4]. Also the efforts that decrease vibration of cylindrical container through the sloshing damping control were introduced. [5, 6] And, the application that vibration control was applied to mobile manipulator robot was proposed. [7, 8]

In this paper, we propose vibration minimization of tower typed 2-wheeled mobile robot using acceleration and deceleration velocity profile method. In the second section, the acceleration and deceleration velocity profile method is described. In the third section, the robotics and experimental environment are explained that would test

* Corresponding author.

Y.-T. Kim, I. Kobayashi, and E. Kim (eds.), *Soft Computing in Advanced Robotics*, Advances in Intelligent Systems and Computing 269,
DOI: 10.1007/978-3-319-05573-2_5, © Springer International Publishing Switzerland 2014

the acceleration and deceleration velocity profile method. In the fourth section, the application of the optimized acceleration and deceleration velocity profile method as to the tower typed 2-wheeled mobile robot through the experiment.

2 Acceleration and Deceleration Velocity Profile Method

Fig.1 show that the rotation Δθ of the wheel of a tower typed 2-wheeled mobile robot makes the disturbance ΔD on the top. Accordingly, the small vibration of wheels can become large vibration on the top.

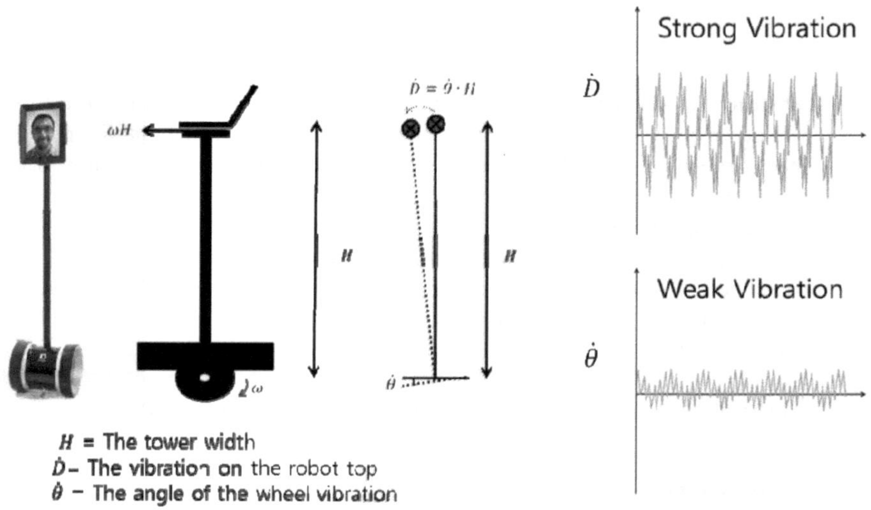

Fig. 1. Tower typed 2- wheeled mobile robot

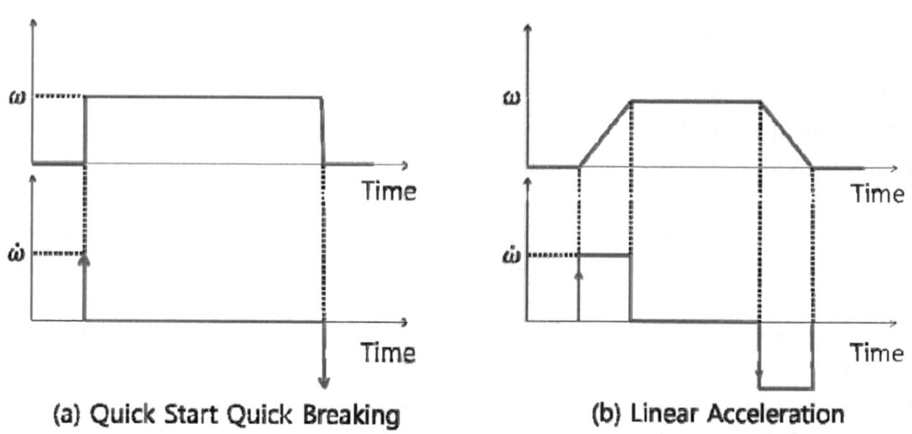

Fig. 2. Angular acceleration according to velocity profile

Fig.2 Show angular acceleration according to velocity profile characteristics. Fig. 2 (a) shows impulse angular accelerations as velocity profile has a quick start and a quick braking. Fig. 2 (b) shows the constant accelerations as velocity has proportional increasing. Fig. 2 (c) shows velocity profile of sigmoid and, acceleration was generated smoothly.

3 Experimental Robot and Experimental Environment

3.1 Experimental Robot

Fig.3 depicts the tower typed 2-wheeled mobile robots that was applied the proposed acceleration and deceleration velocity profile method. The brief specification of on the tower typed 2-wheeled mobile robot is shown as Table1.

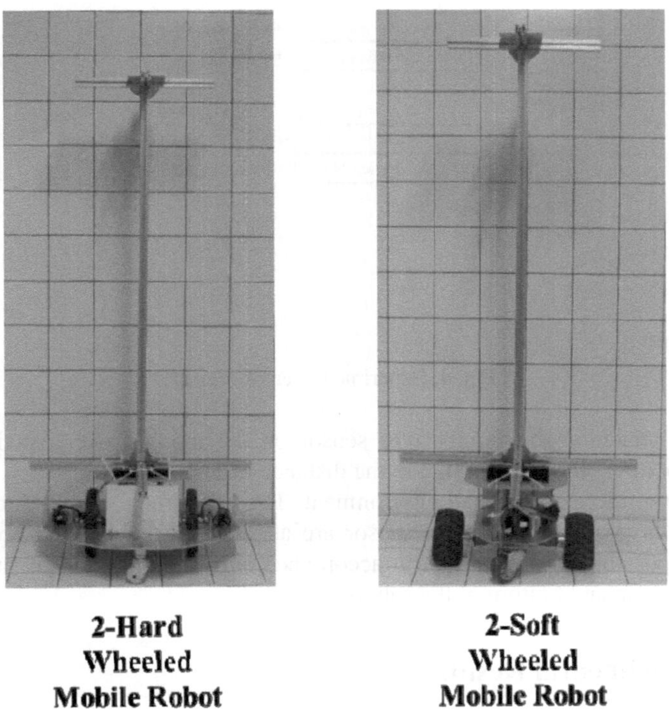

Fig. 3. Experimental Robot

Table 1. Specification of experimental robot

2-Hard Wheeled Mobile Robot	
Criteria	Value
Height	85cm
Weight	9kg
Width	25cm

2-Soft Wheeled Mobile Robot	
Criteria	Value
Height	90cm
Weight	7kg
Width	20cm

3.2 Experimental Environment

As an optimized experimental environment to experiment the acceleration and deceleration velocity profile method in order to reduce the vibration of tower typed 2-wheeled mobile robot, it is to collect the data related to the vibration and acceleration

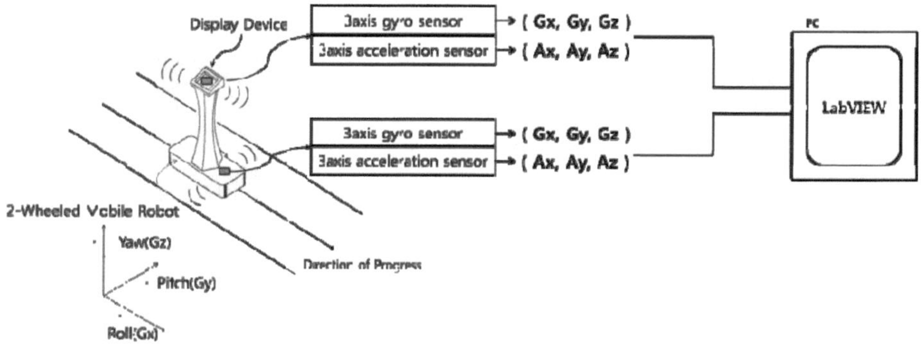

Fig. 4. Experimental environment

velocity by attaching the 3 axis gyro sensor on the top of tower typed 2-wheeled mobile robot along the straight line in the distance of 5M.

Fig.4 shows the experimental environment. The experiment is that three-axis gyro sensor and three-axis acceleration sensor are attached to the top and bottom of the robot. It is possible to obtain data in accordance with the vibration by attaching the sensor on the top and bottom of the robots.

4 Experimental Result

The experiment was conducted after applying the acceleration and deceleration velocity profile method. Shows the vibration by attaching the sensor on the top and bottom of the robots and Show the experimental data. It is possible to find out the difference between the previous data and the current data before the application of acceleration and deceleration velocity profile method through the experimental results. It is possible to know whether the acceleration and deceleration velocity profile method of tower typed 2-wheeled mobile robot works well through the experiment.

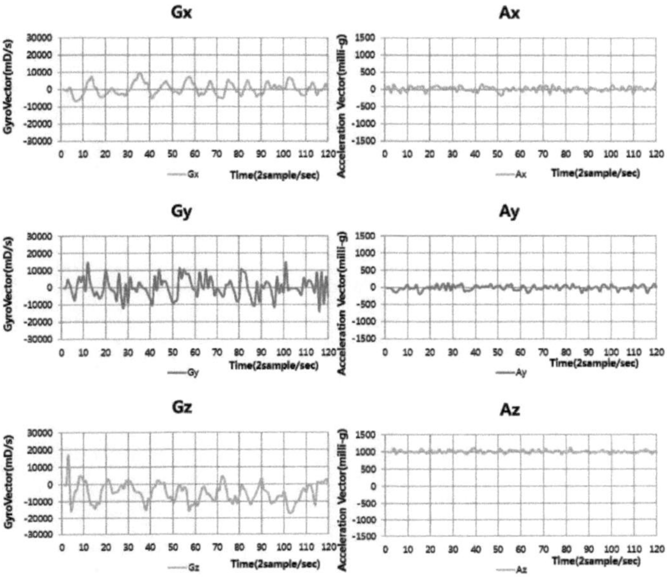

Fig. 5. 2-Soft Wheeled Mobile Robot Quick Start Quick braking (lower part of robot)

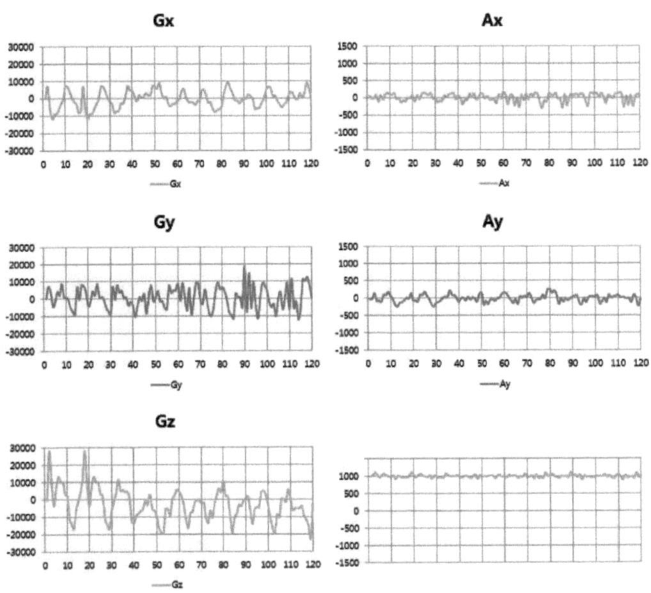

Fig. 6. 2-Soft Wheeled Mobile Robot Quick Start Quick braking (upper part of robot)

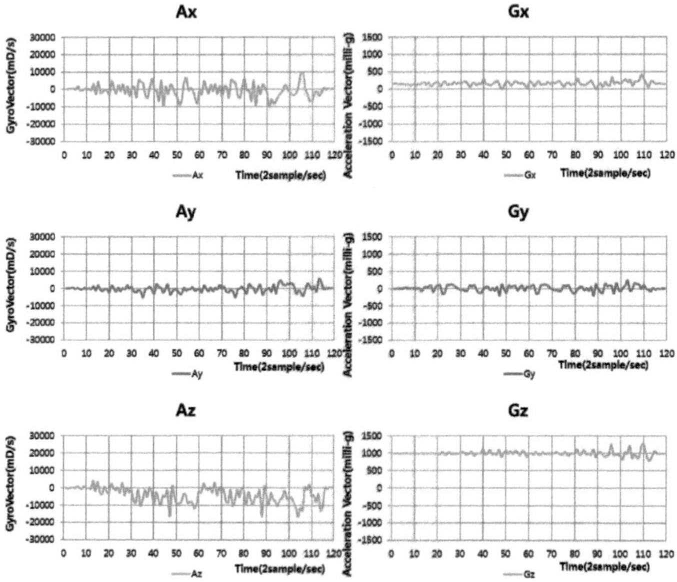

Fig. 7. 2-Soft Wheeled Mobile Robot Smooth Start Smooth braking (lower part of robot)

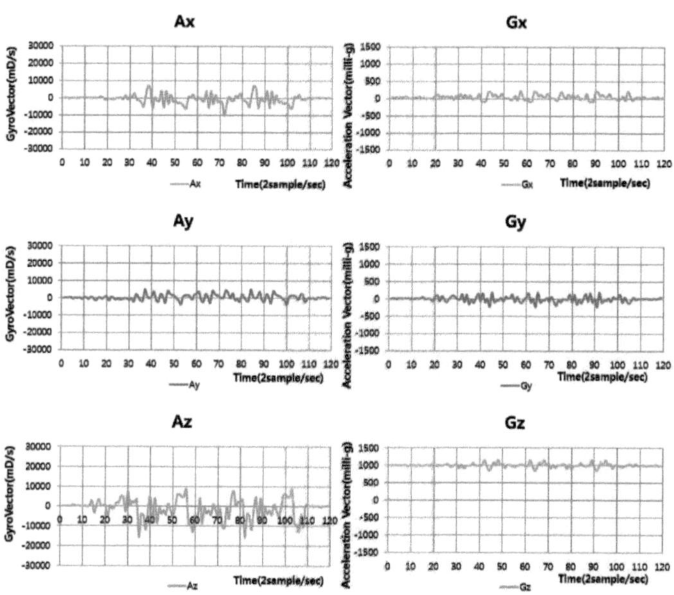

Fig. 8. 2-Soft Wheeled Mobile Robot Smooth Start Smooth braking (upper part of robot)

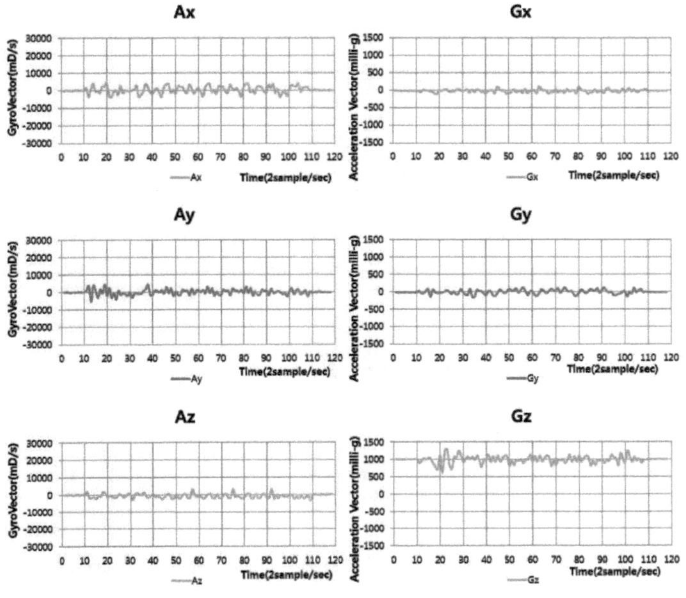

Fig. 9. 2- Hard Wheeled Mobile Robot Quick Start Quick braking (lower part of robot)

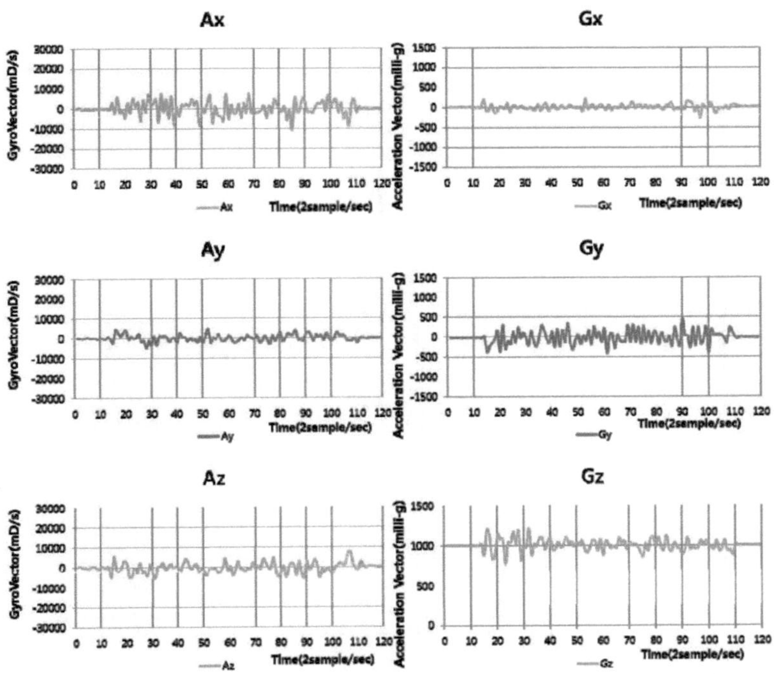

Fig. 10. 2- Hard Wheeled Mobile Robot Quick Start Quick braking (upper part of robot)

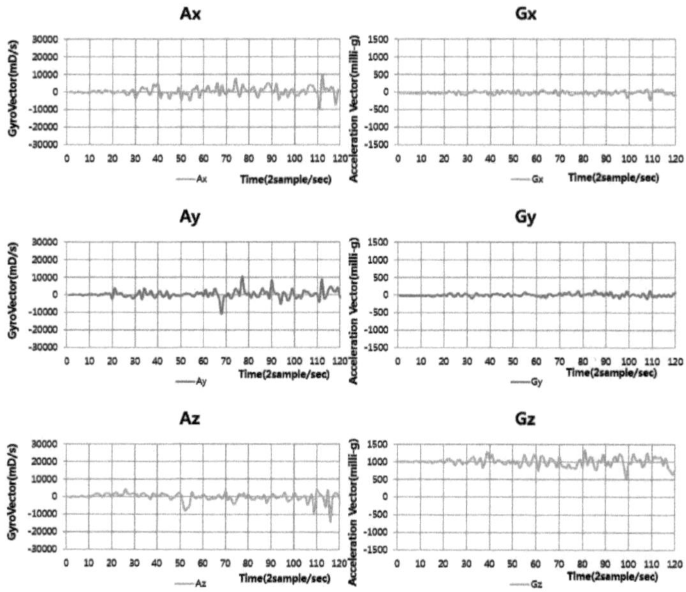

Fig. 11. 2-Hard Wheeled Mobile Robot Smooth Start Smooth braking (lower part of robot)

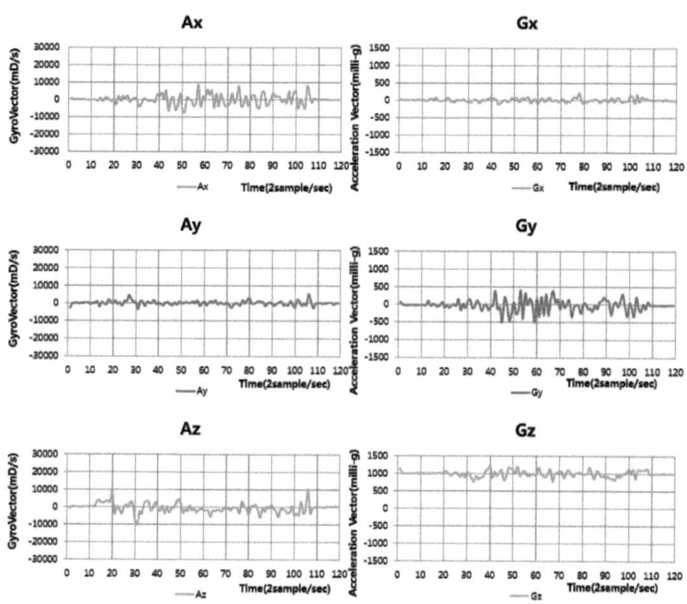

Fig. 12. 2-Hard Wheeled Mobile Robot Smooth Start Smooth braking (upper part of robot)

5 Conclusion

In this paper, we propose vibration minimization of tower typed 2-wheeled mobile robot using acceleration and deceleration velocity profile method. Depicts the tower typed 2-wheeled mobile robot that was applied the proposed acceleration and deceleration velocity profile method. The experiment that could verify the change in vibration was completed by conducting the driving experiment of tower typed 2-wheeled mobile robot as for the acceleration and deceleration velocity profile method that was implemented. Through the experiment, the tower typed 2-wheeled mobile robot optimal acceleration and deceleration velocity profile method have been verified.

Acknowledgements. This research was financially supported by the Ministry of Education, Science Technology(MEST) and National Research Foundation of Korea(NRF) through the Human Resource Training Project for Regional Innovation (No. 2012026068)

References

1. Institute of Electrical and Electronics Engineers, http://spectrum.ieee.org/automaton/robotics/double-robotics-taking-on-telepresence-with-mobile-ipad-base
2. Kato, M., Okugawa, M., Kitagawa, S., Kobayashi, Y.: DSCC2012-MOVIC2012-8727 Attitude Angle Estimation and Vibration Control of Mobile Robot With Self-Righting Mechanism. In: Proceedings of the Asme Dynamic Systems and Control Conference, vol. 3, pp. 113–118. American Society of Mechanical Engineers, New York (2012)
3. Hamaguchi, M., Taniguchi, T.: Damping Control of Liquid Container by Swing-type Active Vibration Reducer on Mobile Robot-Damping Control Method by Using Inverse Model of Sloshing. In: The Society of Instrument and Control Engineers, vol. 47, pp. 346–351. Society of Instrument & Control Engineers (2011)
4. Hamaguchi, M., Taniguchi, T.: Sloshing Damping Control in a Cylindrical Container on a Wheeled Mobile Robot Using Dual-Swing Active-Vibration Reduction. Journal of Robotics and Mechatronics 21, 642–646 (2009)
5. Salehi, M., Vossoughi, G.R.: Vibration Control of Mobile Manipulators Using New Control Element On the End Effector (SMIC/AVEE). In: Proceedings of the International Conference on Advanced Robotics, vol. 13, pp. 578–584. Springer (2007)
6. Mrad, C., Okabe, S., Kamiya, Y., Seki, H.: Vibration Control of Mobile Robot Vehicle by Dynamic Vibration Absorber. JSME International Journal C: Dynamics, Control, Robotics, Design and Manufacturing 42, 62–70 (1999)
7. Mrad, C., Kamiya, Y., Seki, H.: Vibration Control of the End-Effectors of Mobile Rotational Joints Planar Robots. In: Proceedings of the International Conference on Advanced Robotics, vol. 9, pp. 403–408. Japan Robot Association (1999)
8. Nishii, H., Shibatani, K., Suzuki, K.: Vibration Control of a Wheeled Mobile Robot Based on Disturbance Observer. JSME International Journal Vibration, Control Engineering, Engineering for Industry 36, 442 (1993)

A Multi-agent Context-Based Personalized User Preference Profile Construction Approach

Qian Gao[1,2], Su Mei Xi[1,2], Young Im Cho[1,*], and Eric T. Matson[3]

[1] College of Information Technology, The University of Suwon, Hwaseong-si, Korea
gq@qlu.edu.cn, {xsm,ycho}@suwon.ac.kr
[2] School of information, Qilu University of Technology, Jinan, China
[3] Computer and Information Technology, Purdue University, West Lafayette, IN, USA
ematson@purdue.edu

Abstract. This paper proposes a Context-Based Personalized User Preference Profile Construction Approach to comprehensively track the user's local behaviors and user's web behavior of new inputted query. The traditional user profile construction may mainly consider the browsing behavior such as webpage click frequency and webpage click history, but lack consideration of local device context information. So, in this paper, we make use of the context information (interactive historical information and user information that related with the retrieval) which are stored and used in all of the smart devices, owned by the same user, to build and update the user preference profile. Furthermore, in order to avoid the limitation of different vector positions may be allocated to the synonyms of the same term, as well as the size of a document vector must be at least equal to the total number of the words used to write the document, we use the method of ontology-based representation based on WordNet, which uses WordNet to identify WordNet concepts that correspond to the document words. The simulation shows that, our approach can grasp the users' local behavior more accurately, and achieve a higher precision ratio for the method only considering the users' browsing behavior.

Keywords: Multi-Agent, Context, WordNet, Ontology.

1 Introduction

The continuing progress in network technologies and data storage has enabled the digitization and dissemination of huge amounts of documents. Personalized information retrieval and personalized web search are crucial issues in today's world of information overflow, with the main goal of providing only relevant information depending on the context, such that users get the correct information when they need it.

User profiling can be the first step towards personalization of information search and retrieval. Probably, the most frequently used technique for contextualized information-retrieval is relevance feedback [1], although these systems are proven to have a very good performance, but users are often reluctant to add extra interactions with the search engine. So, during the past years researchers have been focusing on

Y.-T. Kim, I. Kobayashi, and E. Kim (eds.), *Soft Computing in Advanced Robotics*,
Advances in Intelligent Systems and Computing 269,
DOI: 10.1007/978-3-319-05573-2_6, © Springer International Publishing Switzerland 2014

implicit feedback techniques [2], where information of the current context is extracted with no need for extra effort from the user. Context-awareness in IR is often based on statistical language models to combine the preceding queries and clicked document summaries. In some way, the documents stored on the user's local smart device can reflect the users' interest, so if we can extract the context information from the user's local smart device, then we can implicitly obtain the users' preference.

The first innovation of this paper is that we make use of the context information (interactive historical information and user information that relates to the retrieval) which are stored and used in all smart devices owned by the same user (such as documents, E-mail, picture as so on) to discern the relevance between the different documents more accurately, based on which the user's preference profile is built. The user's preference profile is updated by using both the context information and the input query, which makes the retrieval reflect the personalized needs more clearly to give precise answers to precise questions.

Second, since the vector space model representation of the document has several limitations, for example, different vector positions may be allocated to the synonyms of the same term. The size of a document vector must be at least equal to the total number of the words used to write the document and so on, so the second innovation of this paper is that we use the method of ontology-based representation based on WordNet, which identifies WordNet concepts that correspond to the document words.

The third innovation of this paper is that we exact the words which have semantic relationships with the input query in the WordNet, based on which the user preference profile is updated.

The rest of this paper is organized as follows: section 2 gives a short review of the related work. Section 3 describes the framework and the realization of the proposed system. Section 4 gives the simulation based on our proposed method and we arrive at the conclusion in Section 5.

2 Related Study

2.1 Semantic Representation Profiles

Semantic network-based profiles are typically built by collecting explicitly positive and/or negative feedback from users. Semantic user profiles can explicitly model the relationship between particular words and higher-level concepts. Thus, they can deal more effectively with the inherent ambiguity and synonymy of natural language. Because the same concept can be expressed with many different words, semantic network profiles in which the nodes represent concepts, rather than individual words, are likely to be more accurate. The SiteIF [3] system builds this type of semantic network-based profile from implicit user feedback. Essentially, the nodes are created by extracting concepts from a large, pre-existing collection of concepts, WordNet [4]. The keywords are mapped into concepts with WordNet. Every node and every arc has a weight that represents the user's level of interest. The weights in the network are periodically reconsidered and possibly lowered, depending on the time of the last update.

InfoWeb [5], a query expansion and document filtering system for an online digital library, also builds user profiles represented as a semantic network of concepts. The profiles are built based on explicit user feedback collected by a browsing agent. It clusters the documents in the collection into a pre-determined set of k possible categories. After the clustering, the document closest to the centroid of the cluster, is selected to act as the representative document for the cluster. As the user interacts with the system, feedback documents are matched to the profile according to a linear combination of the individual terms in the document and the user's semantic network.

In this paper, we combine the method of WordNet and an ontology to represent the document, using WordNet to identify concepts based on the overlap of the local context of the analyzed word with every corresponding WordNet entry. Simultaneously, the ontology-based representation allows the system to use fixed-size document vectors, consisting of one component per base concept.

2.2 Context-Aware Method for User Profiling

Context is an increasingly common notion in Information Retrieval, and has been identified as a major challenge in the field. Several context-sensitive retrieval algorithms exist indeed in the literature, most of which are based on statistical language models to combine the preceding queries and click document summaries with the current query, for better ranking of documents [6]. Relevance feedback, and later, implicit feedback [7], similarly exploits contextual user input as a source of information to complement explicit user queries and guide the retrieval process.

Ph. Mylonas [8] improved the effectiveness of personalization as perceived in a specific context, reducing some of its occasional drawbacks by making it more context-relevant and contextually coherent. They used semantic concepts, rather than terms (i.e. strings), to represent the contextual meanings, and exploited explicit ontology-based information attached to the concepts, so as to automatically extract the persistent semantic user preferences, and live, ad-hoc user interests, which are combined in order to improve the accuracy and reliability of personalization for retrieval.

In this paper we make use of the context information (interactive historical information and user information that related with the retrieval) which are stored and used in all of the smart devices owned by the same user (such as documents, E-mail, picture as so on) to grasp the relevance between the different documents more accurately, so as to capture the users' operation behavior, based on which, we evaluate the importance of different documents.

3 The Proposed Method

3.1 Framework and Workflow of the Multi-Agent Context-Based Personalized User Preference Profile Construction Approach

There are five agents in our entire system, with all agents communicating among each other, and each agent having its own responsibility, as shown in Fig.1.

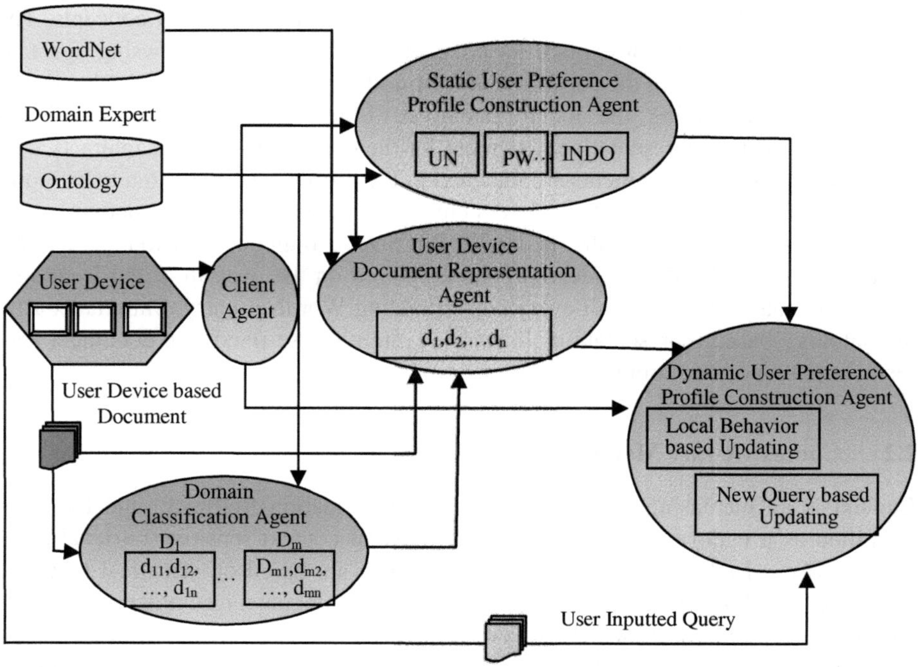

Fig. 1. Context Features based Personalized User Preference Construction Approach for Query Refinement

3.2 Client Agent

The Client Agent is responsible for verifying the user's identity (in the form of "UserID"), calculating the Active Degree (in the form of "UserID.ADj", where *j* is the index of the device) of different user devices, and collect the context information (such as metadata and users' daily working behavior) from all the smart devices owned by a same user.

In order to track the activity of the different user devices, the Client Agent runs a simple loop that periodically sends heartbeat method calls to every device that belongs to user *i*. The heartbeat received from every device tells the device tracker that a device is alive (the heartbeat also doubles as a channel for messages). If the device is alive, then its Active Degree will be set to "+1." This is used to judge the importance of the local behavior by the corresponding device.

The metadata includes file name, creating time, modification time and access time which can be obtained by the win32 API provided by the windows system. And the Local user behavior includes duration information and application program windows switching information which can be obtained by using open source tools "User Activity Logger" [9] to record the user's daily working behavior on the personal computer. Here, we adopt the "Windows Hooks" mechanism and the kernel mode-driven method to acquire the interactive information between users and the application programs.

The User Activity Logger uses log files to record the user's interactive information obtained by the "Windows Hooks" mechanism and kernel mode-driven method.

3.3 Domain Classification Agent

Before we start to perform the document representation, we first prepare the text by performing part-of-speech (POS) [10], Stop of Words Removal, and Words Stemming, and then we use the Domain Classification Algorithm (DCA) to identify which domain the document stored in the user device belongs to. The DCA is based on WordNet domains by comparison between the domain of the document context and the domains of the Word's sense, as shown in Fig.2.

```
Input: Terms in the Documents (D)
Output: The sense belonging to obtained domains is the
correct sense.
Procedure:
1. Get document (D).
2. Perform POS tagging to extract the tagged documents
(D_t).
3. Separate word in (D) according to (D_t)
4. Insert words into bag (B_1) that has count (N)
5. Do
    Insert set of domains corresponding to its POS tag
    into bag (B).
   While i<=N
6. Get target word (W)
7. Insert target word (W) domains corresponding to its
POS tag into bag(B_1).
8. Compare each domain in (B_1) with set of domains of
    remaining content words.
9. If domain in (B_1)>domain of the other content Then
      Set content words are the domain of the text.
      Set sense belongs to domain obtained correct
       sense.
10. End
```

Fig. 2. Domain Classification Algorithm

3.4 User Device Document Representation Agent

Having determined the domain of the document which is stored in the user device belongs to, and then we use the User Device Document Representation Agent to find the weighted corresponding concept in the WordNet to represent the document.

First, using the Concept Identification Algorithm (CIA) to identify which concept in the WordNet is the corresponding one with the words in the document, as shown in Fig.3.

```
Input: all of the words in document D which are gotten
from Words Stemming phase.
Output: Set of WordNet concepts belonging to terms (w-
ords) in documents D.
Procesure: // CTi is the context of the words in the d-
ocuments, it is the sentence  in documents D that con-
tains the word occurrence being analyzed.
1. Do
    Get WordNet entries Ci set(C1,C2,C3,...) that is co-
     ntaining the word Wi.
    Stord Wi in the corresponding Ci in the database
   While i<=N
2. Descending Ranking the concepts Ci according to the
   length of the concept
3. For each Ci
    Get common words between CTi and representative te-
     rms of Ci, which is the intersection CT=∩(CTi,Ci)
     If |CT|<|Ci| then
        The concept sense Ci is not within the context
          CTi.
     If |CT|=|Ci| then
         The concept sense Ci is within the context CTi.
         Add Ci to the set of possible senses associated
with the document
   Endfor
```

Fig. 3. Concept Identification Algorithm

Then, we use the method proposed by Gao and Cho[11] to calculate the weight for each concept of the corresponding document by considering the frequency and the specific of the concept in one document, the generality of the concept in the corresponding domain ontology, and the position of the concept appearing in the document.

3.5 Static User Preference Profile Construction Agent

In the beginning, we initiate the weight for each interested domain a rating scale between 0 and 1 based on the user's interest gathered explicitly during signup. Sum of the weight of all categories within a particular domain will be 1. Initially, there is no term in the user preference profile, as shown in Table 1.

When a user using his/her browser client to first connect to the system or creating an account via the registration page, the proposed system will according his/her inputted information to create the initial static user preference profile

Table 1. User Preference Profile Table

Row Key	User preference with in a particular domain									
	User name	password	Category 1 with its associated weight			…	Category n with its associated weight			
				…		…		…	

3.6 Dynamic User Preference Profile Updating Agent

User interests always constitute the most important part of the user profile in an adaptive information retrieval system that deals with large volumes of information.

A dynamic user profile is used to address frequently changed user interest. The history behavior based updating method is built on the fact that the user behavior based context information on the user device documents (e.g. reading duration information, application programming switching information, file content and metadata, the time spent on a document stored in the users' device, switching frequency and the combination of them) has a strong positive relationship with users' interest. The traditional user profile updating method only considers the users' interactive search session initiated by a particular user on the webpage, and updates the users' interest just depending on the users' browsing behavior and browsing history without considering the users' local documents operating behaviors. As a matter of fact, even though in the offline status, the users' reading behavior on the local file also has a strong positive relationship with users' interest.

Therefore, in this paper, we use two stages to update the user preference profile dynamically. First we update the weight of the each term in the user preference profile based on the users' local behavior. Second, we update the weight of each term in the users' preference profile based on the users' web behavior of new inputted query.

3.6.1 Local Behavior Based Updating Subagent

First, we use the DCA algorithm to classify the user device based document of a particular domain into different categories. For a certain document in a category *j* of domain *i*, by using the user device document presentation agent, it can be presented as $UDP_{ij}(D_k, w) = UDP_{ij}\{(C_{k1},w_1),(C_{k2},w_2),\ldots,(C_{kn}, w_n)\}$. Here $D_k \in$ the documents belongs to domain *i* and category *j*, C_{kt} (t=1,2,…,n) is the concept related to document D_k.

Then, we update the weight of each term in the user preference profile based on the initial weight of each category, the domain document profile, the context information about the user device document to insert.

In order to choose the most suitable items which can best reflect the user's interest, we first use the context information module to collect metadata of the user device document and the user behavior on the user device document, based on which we insert the suitable concepts into the user preference profile and calculate the weight for each concept.

Here we use two kinds of context information (metadata, local user behavior) which have been collected by the Client Agent to weigh the user device document and each concept in it.

①We use Eq. (1) to calculate the time weight of a user device document:

$$CRT(Di) = \alpha \frac{1}{\log(1.6 + \frac{CurrentTime - CreatingTime}{3600})} + \beta \frac{1}{\log(1.5 + \frac{CurrentTime - AccessTime}{3600})} + \gamma \frac{1}{\log(1.4 + \frac{CurrentTime - ModificationTime}{3600})} \quad \text{Here, } \alpha + \beta + \gamma = 1. \tag{1}$$

②Local user behavior includes duration information and application program windows switching information.

We use document-windows switch tree to express the users' application windows switching behavior, as shown in Fig.4.

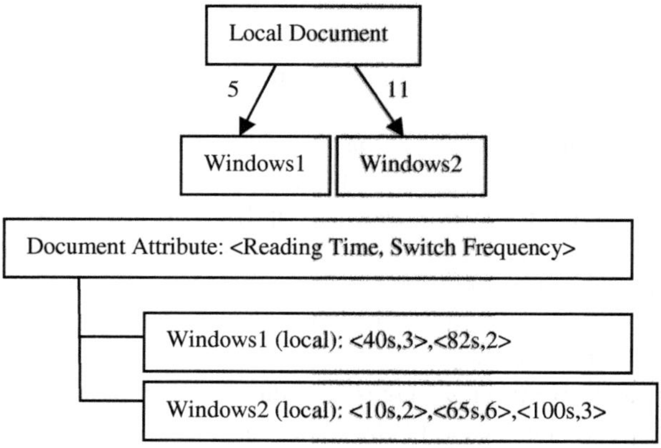

Fig. 4. Document-windows switch tree

The document-windows switch tree is composed of nodes and links. The parental nodes of the first floor indicate the local document window, the child nodes indicate the application program windows which were switched from the local document window and the link indicates the total number of switches. The reading time indicates the time used to browse the local document, during this period, existing click and stroking of the mouse and the keyboard. The switching frequency indicates the total times of switching from the local document windows to other application program windows (local or webpage) sequentially.

When a user browses a document for a period of time, if he sequentially switches to another same window for several times, it means that the content of the two windows is closely related. Furthermore, the total number of times a window has been switched reflects the importance of this document for its user. The more documents have been switched, the more attention of the document got from the user. Considering the lasting time is also an important affecting factor, so we use Eq. (2) to define the switching frequency of document d_i.

$$SF(d_i, CN_i) = \sum_{j=1}^{n} dsw(d_i, window_j, t) \tag{2}$$

Where window belongs to the child node set of d_i in the document-window switch tree, $dsw(d_i, window_j, t)$ indicates the switching frequency of document d_i directly switching to the child node $windows_j$ in the period of t, t is a constant that can be decided by the user, and $SF(d_i, CN_i)$ indicates the switch frequency of document d_i directly switching to all of its child nodes. For example, for Fig.4, $dsw(d_i, window_1, t)=3+2=5$.

As previously mentioned, the more the document has been switched the more attention of the document from the user. So, first we use the switch frequency to calculate the similarity of two documents, as shown in Eq.(3), and based on it to calculate the behavior weight of the document so as to reflect how important the document will be, as shown in Eq.(4).

$$Sim(d_i, d_j) = \frac{dsw(d_i, d_j, t)}{RT(d_i, d_j, t)} \tag{3}$$

Where RT (d_i, d_j, t) indicates the total reading time before the d_i was switched to d_j in the period of t.

$$BWweight(d_i) = \frac{1}{n} \sum_{j=1}^{n} Sim(d_i, d_j) \times SF(d_i, CN_j) \tag{4}$$

Where n indicates the total number of child nodes, d_j belongs to the child nodes.

Then we combine the Eq.(1) and Eq.(4) to get the local behavior based weight of the document i, as shown in Eq.(5).

$$LBW(di) = a \times BWweight(d_i) + b \times CRT(d_i) \tag{5}$$

Where a and b are two constants satisfying a+b=1, and the value of a and b will be obtained by the simulation.

Furthermore, we use the method proposed by Gao and Cho[12] to calculate the weight of the concept appearing in the user device document, UDW $(d_i)=\{(c_{i1},w_1),(c_{i2},w_2),\ldots\}$, which is combined with the $LBW(d_i)$ to calculate the weight of each concept in $document_i$, as shown in Eq.(6).

$$LW(d_i)=LBW(d_i) \times UDW(d_i) = LBW(d_i) \times \{(c_{i1},w_1),(c_{i2},w_2),\ldots\} \tag{6}$$

Since the activity of the smart device can also affect the importance of the document (That is to say, if the smart device *i* is rarely to be used, then the documents stored on this device is less important than the documents stored on the other frequent used smart devices), so finally, we combine the Active Degree which we have got by the Client Agent to refine the weight of each concept in $document_i$, as shown in Eq.(7).

$$RLW(d_i) = \sum_{i \in smart\ device} Actice\ \ Degree \times LW(d_i) \tag{7}$$

Here, $RLW(d_i)$ is a vector space model that represent the weight of each concept in document i.

The algorithm of the local behavior based user profile updating is shown as Fig.5.

3.6.2 New Query Based Updating Subagent

First, whenever a user clicks a new webpage, we update the weight of each concept in the user profile with the method proposed by Gao and Cho [13] based on the generality of the concept C which appeared in the newly inserted documents.

Second, we update the user preference profile based on the users' newly input query by considering semantic similarity between the input query and the synonym concept, hyponym concept, and meronym concept that can be extracted from WordNet.

Since the new term will be added constantly into the user preference profile, the scale of the user preference profile will become bigger and bigger. As a result, after a period of time, some old terms must be removed in order to reduce the users' preference profile. Here, we use Eq. (8) to calculate the degree of similarity of two concepts:

$$S(C_i, C_j) = -\log \frac{len(C_i, C_j) + 1}{2 \times depth} \tag{8}$$

Where, $len(C_i, C_j)$ makes reference to the length of the shortest path between the synsets; depth is the height of the classifier tree. When two words have the same meaning $len(C_i, C_j) = 0$, depth = 1, $S(C_i, C_j)=1$.

The update of the new query based method is shown in the algorithm of Fig.6.

4 Simulation Result

In order to perform an evaluation of the accuracy of the search with a user profile, it was necessary to develop a test client displaying the results of a search. We constructed an HDFS Cluster to simulate our proposed method (a single computer with an Intel Core2 CPU, 4 GB RAM as NameNode, and 2 computers with Intel Core2 CPUs and 4 GB RAM as DateNode). We used Java as the programming language and the Eclipse integrated development environment. In this study, we analyzed and operated the ontology files using an open-source framework "Jena."

To test the approach proposed in this paper, we adopted the TREC 10 query sets as the tools, and eight Ph.D. and Masters students from different fields of computer science and education participated in the experiments.

```
Input: Content based information acquired by Win 32
API and User Activity Logger.
Output: Local behavior based user profile
Procedure:
1. Do while i<=n    //n is the number of the document
      using WSD algonthm to identify which the domain
      that document belongs to.
   End Do
2. Do While j<=m     // m is the number of the domain
   using WSD algonthm to identify which category the
   document c_j belong to.
   Enddo
3. Do While i<=m     // m is the number of the domain
    Do While j<=c    // c is the number of category in
                        domain i
     For document k belongs to domain i of category j
        Use eq(1) to caculate the time weight of the
        user device documents
      Use eq(5) to caculate the local behavior based
        weight of device documents
      Use eq(6) to caculate the weight of each conce-
        pt in document k
       Use eq(7) to recaculate the weight of each con-
        cept in document k
      Choose the top 10 concepts of document k
     Endfor
     Do while t <= 10
       If concept C_kt is not in the user preference
         profile then
           insert( C_kt, W_kt )into the profile
       Else
         W_t = W_t + W_kt// W_t is the weight of concept C_t
      Enddo
      Normalize the weight of each concept in the us-
       er profile as W_t = W_t / the total number of docu-
       ment belongs to domain i of category j
    Enddo
   Enddo
4. If a user creates and browses a new document Then
     repeat using WSD algorithm to idendify the domain
    and category.
5. Goto step3
6. End
```

Fig. 5. Local behavior based user profile updating algorithm

```
Input: New Query
Output: Updated weight of the concept in the user pro-
        file
Procedure:
For each item Ci in the new query
   If the newly input query already exists in the user
    preference profile then
       weight=weight+3;
   Else
     If the input query has synonym Cj in the user pre-
     ference profile then
       weight=weight+S(Ci,Cj);
     Else
       If the input query has hyponym Ck in the user
        preference profile then
          weight=weight+S(Ci,Ck);
       Else
         If the input query has meronym Ct in the user
         preference profile then
           weight=weight+S(Ci,Ct);
       If the weight of the concept Ci is less than the
         threshold T then
          Remove Ci out of the user preference profile
EndFor.
```

Fig. 6. New query based updating algorithm

Each series of test requests a search based on a set of dynamic updated user profiles: an English speaker with text as his displayable format. A comparison is made between a search request making use of a profile and one with only semantic web based method which uses traditional vector space model (VSM), also called term frequency inverse document frequency to carry out searching, and evaluate the performance of the two methods from the precision and recall, defined as Eq.(9) and (10).

$$Precision\ \ Ratio = \frac{\#relevant\ \ items\ \ retrieved}{\#retrieved\ \ items} = P(relevant \mid retrieved) \tag{9}$$

$$Recall\ \ Ratio = \frac{\#relevant\ \ items\ \ retrieved}{\#relevant\ \ items} = P(retrieved \mid relevant) \tag{10}$$

The simulation results are shown in Figs. 7, 8.

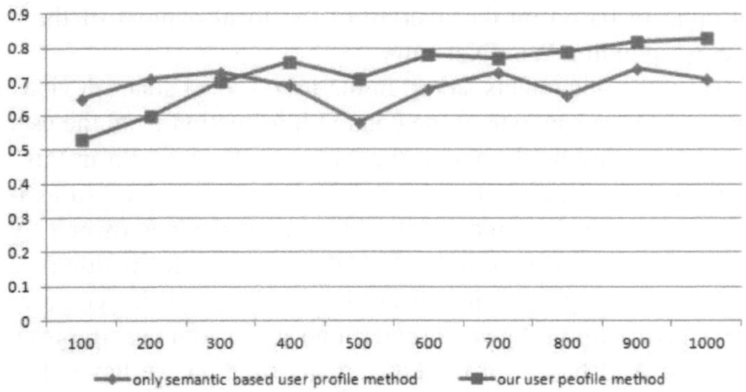

Fig. 7. Precision Ratio of only semantic based user profile method versus our proposed method

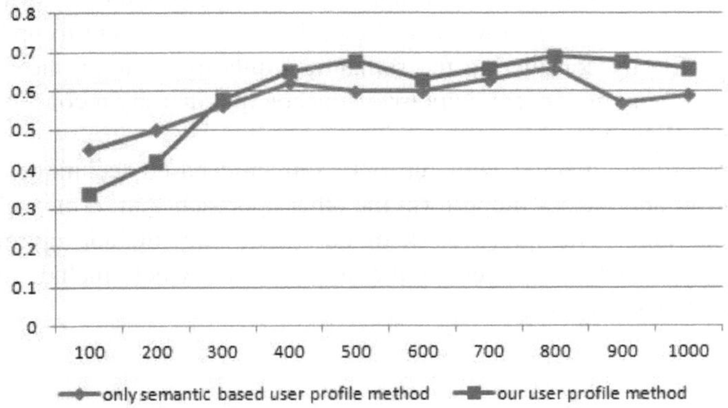

Fig. 8. Recall Ratio of only semantic based user profile method versus our proposed method

The overall results show that at the beginning, since the user profile is not perfect, so the Precision Ratio and the Recall Ratio of our proposed system is lower than the only semantic based retrieval system. Along with the gradual perfection of the user profile, more retrieval results will be selected by our system and more users' interests will be considered, so the performance of our approach is better than only semantic based user profile method. In addition, since the users' long-term interests may change with time and the short-term interests may occur (maybe only for once retrieval), there might be a fluctuation on Recall Ratio and Precision Ratio before the users' profile is perfect. But once the user profile is perfect, the Recall Ratio and Precision Ratio will be maintained on a steady level

5 Conclusion

In this paper, we combine the method of WordNet and ontology to represent the document. In User Device Document Representation Agent, we use WordNet to identify a WordNet concept based on the overlap of the local context of the analyzed word with every corresponding WordNet entry.

In some ways, the documents stored in the user's local smart device can reflect the user's interest, so if we can extract the context information from the user's local smart device, then we can implicitly obtain the user's preference. So in this paper, we use Client Agent to track the activity of the different user devices so as to evaluate the importance a certain device among all of the smart devices owned by a same user, meanwhile we use it to monitor and collect the context information (such as metadata and users' daily working behavior) from the user's smart devices owned. Then, we use Domain Classification Agent to determine which domain the user device documents belong to. And based on the context information we have obtained from the Client Agent we using Local Behavior Based Updating Subagent to update the user preference profile by comprehensively considering the metadata and local user behavior such as duration information and application program windows switching information. Finally, we use the New Query based Updating Subagent to update the user preference profile by considering the semantic similarity between the inputted new query and the synonym concept, hyponym concept, and meronym concept that can be extracted from WordNet.

We compared our method with the only semantic based technique. The results showed that our approaches outperform the other approaches in terms of Recall Ratio and Precision Ratio. The future work of our paper may include optimization and adaptability of dynamic changes to the user profile which include the frequent changes with user's interests.

References

1. Rocchio, J., Salton, G.: Relevance feedback in information retrieval. Prentice-Hall (1971)
2. Kelly, D., Teevan, J.: Implicit feedback for inferring user preference: a bibliography. SIGIR Forum 37(2), 18–28 (2003)
3. Stefani, A., Strappavara, C.: Personalizing Access to Web Sites: The SiteIF Project. In: Proceedings of the 2nd Workshop on Adaptive Hypertext and Hypermedia, HYPERTEXT 1998, June 20-24 (1998), http://www.contrib.andrew.cmu.edu/~plb/HT98_workshop/Stefani/Stefani.html
4. The Wordnet Website, http://wordnet.princeton.edu/
5. Gentili, G., Micarelli, A., Sciarrone, F.: Infoweb: An Adaptive Information Filtering System for the Cultural Heritage Domain. Applied Artificial Intelligence 17(8-9), 715–744 (2003)
6. Finkelstein, L., Gabrilovich, E., Matias, Y., Rivlin, E., Solan, Z., Wolfman, G., Ruppin, E.: Placing Search in Context: The Concept Revisited. ACM Transactions on Information Systems 20(1), 116–131 (2002)
7. Shen, X., Tan, B., Zhai, C.: Context-sensitive information retrieval using implicit feedback. In: Proc. of the 28th Annual International ACM SIGIR Conference on Research and Development in Information Retrieval, SIGIR 2005, Salvador, Brazil, pp. 43–50 (2005)
8. Mylonas, P., Vallet, D., Castells, P.: Personalized information retrieval based on context and ontological knowledge. The Knowledge Engineering Review 23, 73–100 (2008)

9. Chernov, S., Demartini, G., Herder, E., et al.: Evaluating personal information management using an activity logs enriched desktop dataset. In: Florence Italy Proceedings of 3rd Personal Information Management Workshop (2008)
10. The Stanford Natural Language Processing Group, `http://nlp.stanford.edu/software/tagger.shtml`
11. Gao, Q., Cho, Y.I.: A Multi-Agent Personalized Ontology Profile Based Query Refinement Approach for Information Retrieval. In: 2013 13th International Conference on Control, Automation and System (2013) (in press)
12. Gao, Q., Xi, S.M., Cho, Y.I.: A Multi-Agent Personalized Ontology Profile Based User Preference Profile Construction Method. In: 2013 44th International Symposium on Robotics (2013) (in press)
13. Gao, Q., Cho, Y.I.: A Dynamic Ontology-based Multi-Agent Context-Awareness User Profile Construction Method for Personalized Information Retrieval. International Journal of Fuzzy Logic and Intelligent Systems 12(4), 270–276 (2012)

Fuzzy Visual Navigation Method for Autonomous Freight Transportation Robot

Dong-Hyuk Jeong[1], Jong-Hwan Yoon[2], and Yong-Tae Kim[1,2,*]

[1] Dept. of Electrical, Electronic and Control Engineering, Hankyong National University
327 Joongang-ro, Anseong, Gyeonggi-do, 456-749, Korea
[2] Smart Logistics Technology Institute, Hankyong National University
167 Joongang-ro, Anseong, Gyeonggi-do, 456-749, Korea
{jdhyuk87,neosanson,ytkim}@hknu.ac.kr

Abstract. In this paper, we proposed a fuzzy visual navigation method for using RFID-based map. At first, we design a virtual map which is composed of RFID node and generate a path for navigation. If the robot recognizes a pallet while driving, it uses a proposed fuzzy visual navigation. While the robot is driving to carry out transportation function in the various environments, it uses sensor information from ultrasonic sensors, RFID sensors and camera. We design a freight transportation robot in order to move a goal position effectively. The proposed methods are verified by a navigation experiment.

Keywords: fuzzy visual navigation, autonomous navigation, RFID map, freight transportation, logistics robot.

1 Introduction

Many researchers have been interested in automation of logistics process and have expected that IT technology can be applied to logistics industry. Key challenges to wide-spread deployment of robot are collaboration and cooperation ability to adapt sensing and information processing to achieve tasks. Therefore, the robots are increasingly being deployed in the real world application domains such as manufacturing factory and service industry due to the application availability of high-fidelity sensors and sophisticated intelligent algorithms to process sensor information. According to various demands of consumers and reduced cycle of product handling, the logistics automation technology is essential for efficient logistics [1-3].

Automatic guided vehicle(AGV) system is recently used in various applications field such as transportation, material handling, planetary exploration, mining, military operation, etc. Also AGV systems are widely used to increase the flexibility and the efficiency of the material handling systems. As electronics and information technology are advanced, so did AGV applications. The vehicles become more intelligent and are spread in various fields. There are several hundred AGV systems in operation.

Especially, AGV system is a critical factor which determines the overall performance of the manufacturing systems. AGV needs a path of generation and

Y.-T. Kim, I. Kobayashi, and E. Kim (eds.), *Soft Computing in Advanced Robotics*, Advances in Intelligent Systems and Computing 269,
DOI: 10.1007/978-3-319-05573-2_7,

tracking, real-time self-localization for the distribute center [4][5]. There are several previous navigation methods using magnet-gyro guidance and wire guidance. But these methods have disadvantages such that it is hard to change the environment flexibly and needs a high cost to keep the maintenance of working environment [6]. To solve these problems, RFID system can be used to generate the navigation map to compensate location error [7] and visual navigation method can be designed for the autonomous navigation and target tracking of the transportation robot [8].

In this paper, we design an autonomous freight transportation robot and propose operation method of the robot using RFID system and fuzzy visual navigation. We verify the proposed method through experiments using a designed freight transportation robot. We show that the freight transportation robot recognizes a pallet calibrating direction and moves to target while uplifting a pallet. We describe the calibrating direction structure and the mechanism of freight transportation robot in section 2. In section 3, we propose the operation method for freight transportation robot using RFID map and visual navigation method. In section 4, we present experimental results. Finally, we conclude with a discussion in Section 5.

2 Freight Transportation Robot System Construction

2.1 Robot System Structure

Fig. 1 shows the overall architecture of freight transportation robot that comprises a main controller, a sub motion controller, a user PC, 2 BLDC motors, lift module and battery. The main controller is an ARM CORTEX-A9 embedded board using Linux operating system. The embedded board controller performs on-board image processing, motion planning, navigation algorithm processing and LRF sensing. The motion controller is an ARM CORTEX-M3 processor board that interfaces with a sensor module. It also controls 2 BLDC in real time. The sensor controller is composed of a RFID sensor and 12 Ultrasonic sensors. The battery was used to supply 24V power for motors and controllers.

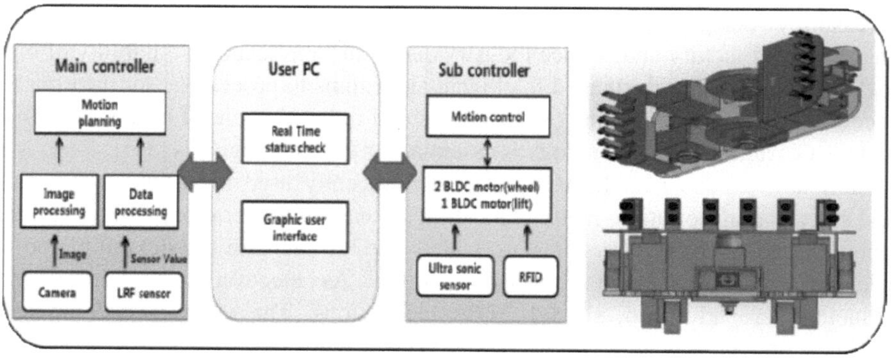

Fig. 1. Overall structure of autonomous freight transportation robot

2.2 Kinematic Model of Freight Transportation Robot

In the general case, steering system of the mobile robot can be classified as differential drive, three-wheel drives and multiple-wheel drive. Differential drive system changes the direction using velocity difference between both wheels. It comparatively has a simple mechanism, so that it is frequently used for many applications [9].

In the paper, we apply a differential drive method to the robot and use 2 BLDC motors in the wheels. The lift module of robot is designed to get on and off a loaded baggage through up and down motion. We use ultrasonic sensor, LRF sensor, RFID, and camera sensor to carry out transportation function while the robot is driving in the various environments. The mobile robot is designed to have a structure to be arrived at goal position effectively.

Kinematic model of the freight transportation robot is shown in Fig. 2. (x_c, y_c) is a coordinate of center position of the robot and v is a robot linear velocity. V_L and V_R are linear velocity of both wheels. θ is the directional angle toward front direction. To obtain a simple kinematics model, we presume that robot wheels are satisfied with non-slipping condition. According to this condition, we can obtain Eq. (1) and Eq. (2). Also, L means distance between both wheels.

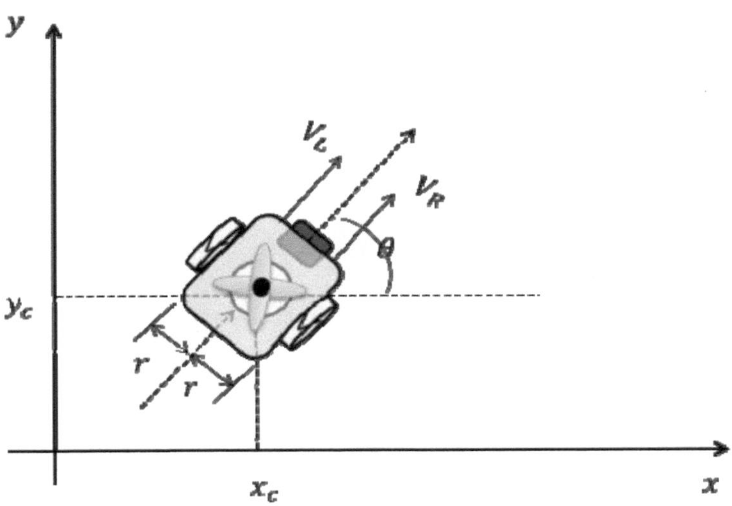

Fig. 2. Kinematic model of the freight transportation robot

$$w = \frac{V_R - V_L}{L} = r\frac{W_R - W_L}{L} \tag{1}$$

$$v = \frac{V_R + V_L}{2} = r\frac{V_R + V_L}{2} \tag{2}$$

3 Operation Method of the Freight Transportation Robot Using Fuzzy Visual Navigation

3.1 Autonomous Navigation Method of the Freight Transportation Robot

For autonomous navigation at the warehouse, the robot needs a path from start position to goal position. Also, the robot has to track this path to the target position. Therefore, it always has to recognize a present location and consider location compensation for the navigation in the real environment. Conventional localization methods use a information of motor encoder but error accumulation phenomenon occurs by skidding on the wheel. Many previous methods are proposed to solve these problems.

In this paper, we apply RFID and QR-code system for the navigation environment. Fig. 3 represent an overall operation sequence. At first, we generate a virtual map which is composed of nodes. Each node contains the coordinate information of the environment. A navigation path from start position to goal position is generated by using A* algorithm. Fig. 4 represents a virtual map of the workspace. If autonomous navigation of the transportation robot starts, it moves from present node P_n to next node P_{n+1} along the generated path and recognizes each node in real time processing. Also, the robot checks the coordination information in the each node and tracks the path without deviation from the path.

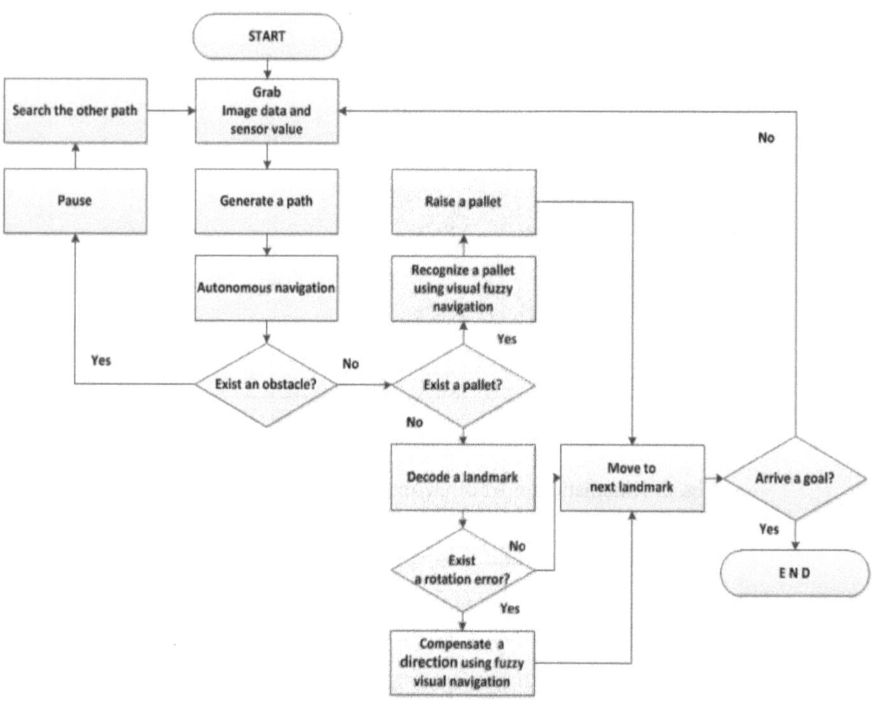

Fig. 3. The operation sequence of freight transportation robot

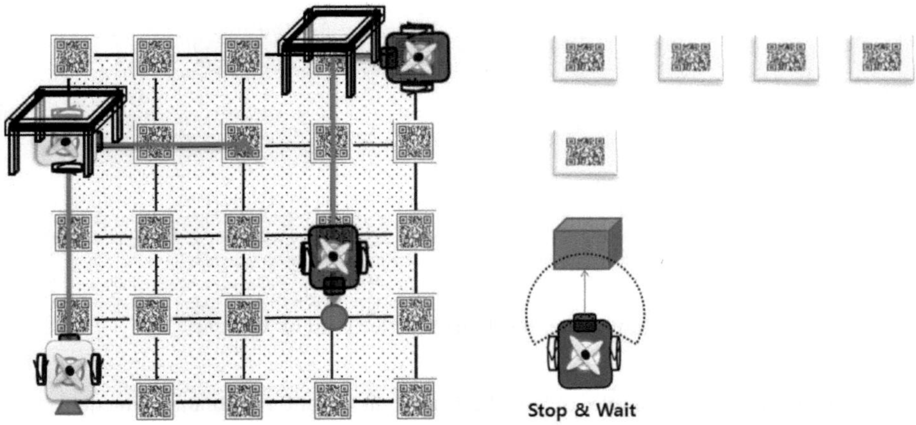

Fig. 4. Virtual map of workplace based on RFID and QR-Code node

When the robot moves to goal position, we have to consider about tracking the path without rotation error. So we propose a method of calibrating direction using visual fuzzy navigation. If the robot decodes the QR-code on the each node, we can get rotation error θ_r and center point P like a fig. 5. Fig.6 and table 1 represent fuzzy rules for calibrating direction. In the camera view, there are off-set area and 5-types separated area. If the x_c of point P is large in the offset area (PB), the robot drives to draw curve deeply. If the θ_r is large (PB) in the offset area, the robot needs a lot of rotation speed of its body.

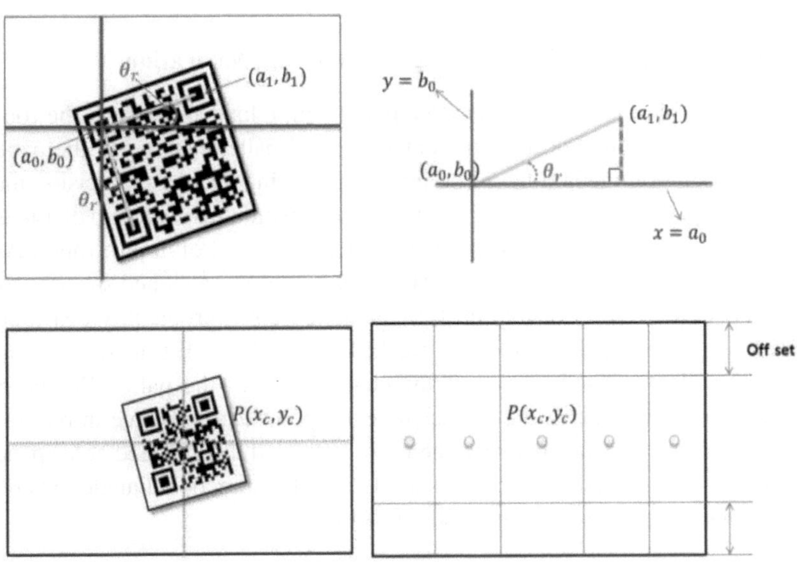

Fig. 5. The method of calibrating rotation error

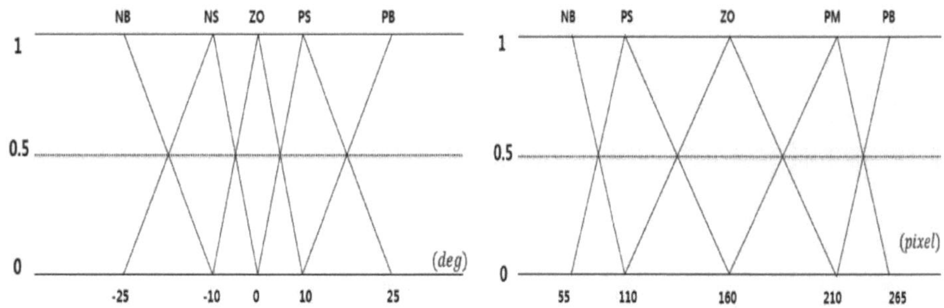

Fig. 6. Membership function of θ_r(left) and of x_c(right)

Table 1. Fuzzy control rules of left wheel and right wheel for calibrating direction

x_c \ θ_r	NB	NS	ZO	PS	PB
NB	PM/ZO	PS/ZO	PB/PM	ZO/PS	ZO/PM
NS	PM/ZO	PS/ZO	PM/PS	ZO/PS	ZO/PM
ZO	PM/ZO	PS/ZO	PB/PB	ZO/PS	ZO/PM
PS	ZO/PM	ZO/PS	PS/PM	ZO/PS	ZO/PM
PB	ZO/PM	ZO/PS	PM/PB	ZO/PS	ZO/PM

3.2 Pallet Recognition Method Using Fuzzy Visual Navigation

In order for the robot to drive and move a pallet with a loaded luggage, the robot has to find a pallet position and move to center point of pallet exactly. In the paper, we propose a recognition method based on image processing and fuzzy decision method. Fig. 7 shows a pallet recognition algorithm. At the position just before pallet node, the robot starts image processing using camera which is located on top of robot. At first, the grabbed RGB image is converted to HSV image and pre-attached specific color of the pallet is set as ROI. A histogram for calculating color probability distribution is calculated. Then designated color is tracked using Camshift algorithm and the algorithm can also be used to extract the center point of the pallet. We propose a visual fuzzy navigation using the extracted center point of coordinate in real time. In the Fig. 8, d means distance between the robot body and the attached color paper and 4 kinds of fuzzy membership function is used. θ is the directional angle toward front direction. And it uses 5 fuzzy membership functions.

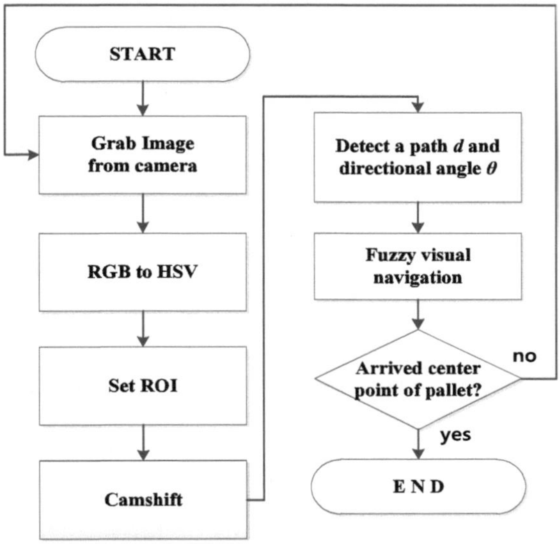

Fig. 7. Pallet recognition algorithm based on image and fuzzy decision

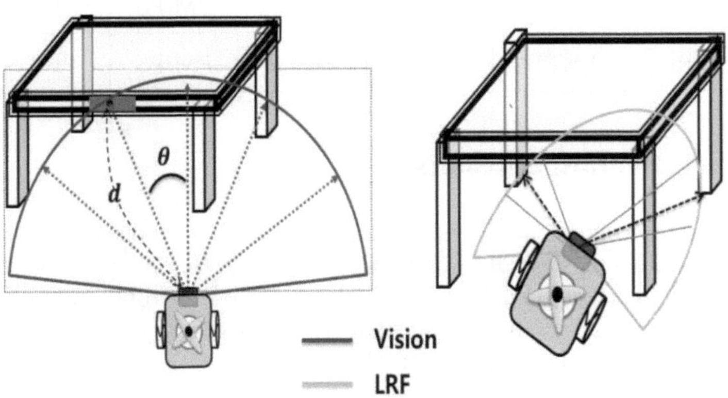

Fig. 8. Picture of pallet recognition method

Fig. 9 presents fuzzy membership functions and Table 2 represents fuzzy control rules for visual navigation method. We designed a fuzzy controller to control direction and velocity depends on direction and distance value. If d is large (PB), the robot move fast. On the other hand, if d is small (PS), the robot move slowly. If θ is large (PB), the robot needs a lot of rotation of its body. The fuzzy visual navigation repeats again until tracking window is less than the set-value. The robot goes ahead while detecting the legs of pallet using LRF sensor information and arrives under the center point of the pallet. After robot is arrived at the center point, it stops movement and operates the lift module for raising the pallet. And then, the robot moves again toward the goal position.

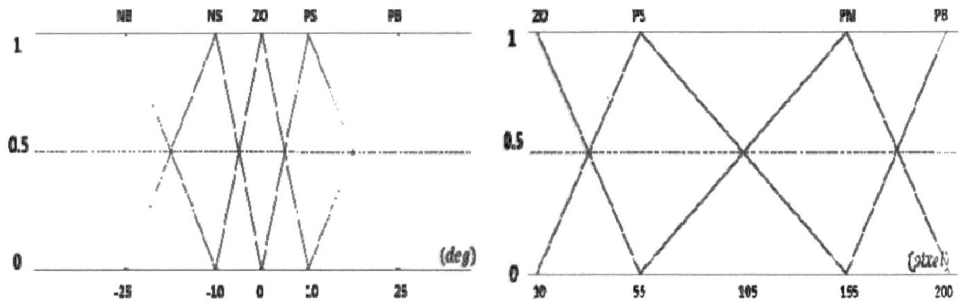

Fig. 9. Membership function of θ(left) and of d(right)

Table 2. Fuzzy control rules of left wheel and right wheel for tracking pallet

d \ θ	NB	NS	ZO	PS	PB
ZO	NM/NM	NS/NS	ZO/ZO	NS/NS	NM/NM
PS	NS/NM	ZO/PS	PS/PS	PS/ZO	NS/NB
PM	PS/PM	PS/PM	PM/PM	PM/PS	PM/PS
PB	PS/PB	PS/PM	PB/PB	PM/PS	PB/PS

4 Experimentation Results

We verify the proposed method by using a designed freight transportation robot. Fig. 10 shows a GUI snapshot of monitoring program. We develop the monitoring program with network communication and check a status of robot in real time. Also, the user can command to the robot such as an emergency stop, simple control commands by using a monitoring program.

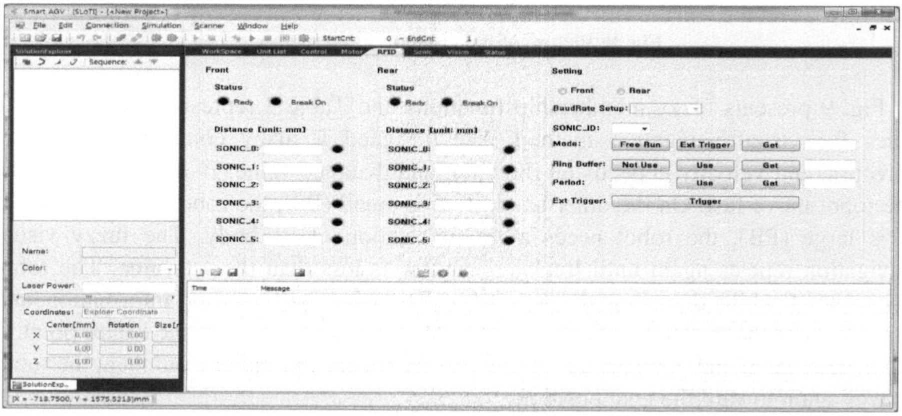

Fig. 10. A GUI snapshot of the monitoring program

Fig. 11 represents an experimental result of rotation error compensation. We verified that obtain decoded coordination information and calculate rotation error through proposed method.

Fig. 11. Snapshots of rotation error calibration

To verify performance of pallet recognition method, we had experiment according to the distance from pallet to robot and rotation angle. Table. 3 and Fig. 12 are shown that success rate of the task accomplishment is about 97% on average.

Table 3. Acheivement of autonomous navigation experiments

d \ θ	-30	-15	0	15	30
1m	9/10	9/10	10/10	10/10	8/10
2m	10/10	10/10	10/10	10/10	10/10
3m	10/10	10/10	10/10	10/10	10/10

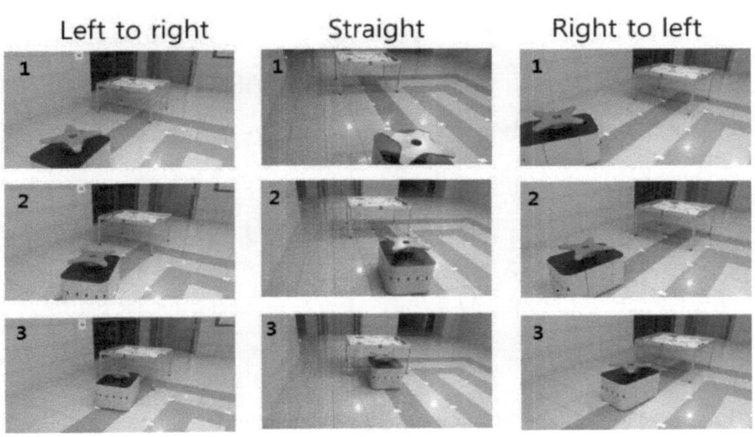

Fig. 12. A snapshots of pallet recognition experiment

Fig. 13 represented a path generation for navigation from main PC controller. Green box means start position, blue boxes are the obstacle or terrain features in the distribute center and red box is target position. In the bottom left side, there is a list of robot moving sequence which is composed of coordination information.

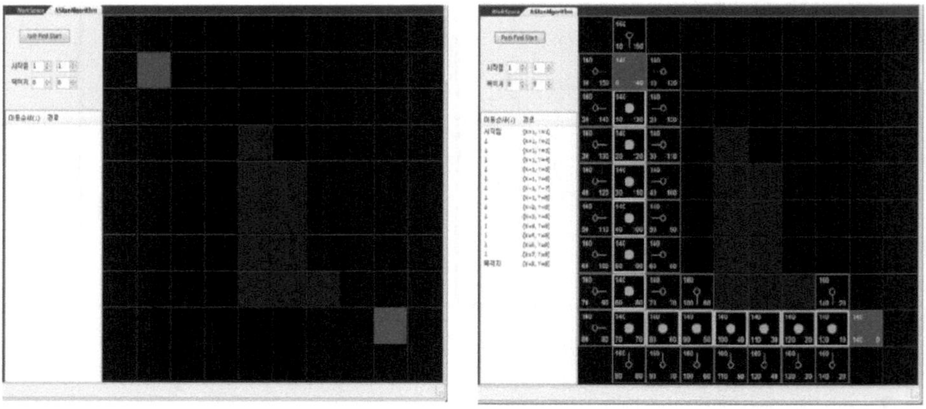

Fig. 13. Path generation of freight transportation robot

Fig. 14 shows the snapshots of freight transportation and autonomous driving experiments using the developed robot. If the freight transportation robot generates a path, the robot tracks the path in order to move a pallet to a goal position.

As result of navigation experiments, it is shown that the freight transportation robot successfully moves to the target while uplifting a pallet. We also verified the performance of the proposed operation autonomous navigation method.

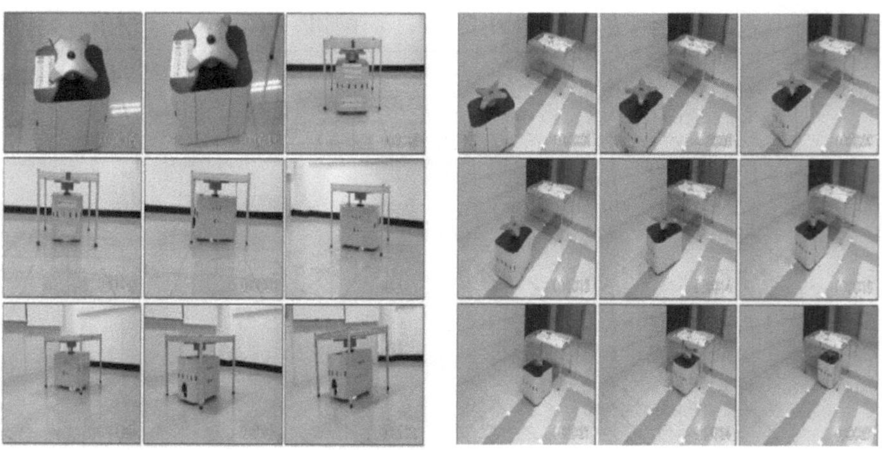

Fig. 14. Snapshots of the navigation experiment using the freight transportation robot

5 Conclusion

In this paper, we design a freight transportation robot and propose operation method of the robot. At first, we design an autonomous freight transportation robot that has the lift module for baggage loading and uses a differential drive for the effective movement and rotation. We propose a virtual map which is composed of RFID node and a path planning method using the map. The robot moves under the target pallet by using fuzzy visual navigation method. The experimental results demonstrate that the proposed robot system can be applied to freight transportation tasks in the warehouse. In the future, we will study optimal path planning and obstacle avoidance in the dynamic environment.

Acknowledgements. This research was supported by the Basic research Program for General Researchers through the National Research Foundation of Korea(NRF) funded by the Ministry of Science, ICT and Future Planning (No.2010-0023291).

References

1. Vis, I.A.: Survey of Research in the Design and control of Automated Guided Vehicle System. European Journal of Operational Research 170(3), 677–709 (2006)
2. Schulze, L., Wullner, A.: The Approach of Automated Guided Vehicle Systems. In: IEEE International Service Operations and Logistics, and Informatics, pp. 522–527 (2006)
3. Amazon.com, Inc., http://www.kivasystems.com
4. Kim, J.M., Park, J.J., Jeon, T.R., Kim, S.S.: Fuzzy and Proportional Controls for Driving Control of Forklift AGV. Journal of Korean Institute of Intelligent Systems 19(5), 699–705 (2009)
5. Ko, N.Y., Seo, D.J., Moon, Y.S.: A Method for Real Time Target Following of a Mobile Robot Using Heading and Distance Information. Journal of Korean Institute of Intelligent Systems 18(5), 624–631 (2008)
6. Hong, H.J., Ro, Y.S., Kang, H.J., Suh, Y.S., Kim, T.H.: Development of the Simulation Tool for The Modeling and Traffic Control of a AGV System. Journal of Korean Institute of Intelligent Systems 14(4), 499–505 (2004)
7. Park, J.H., Kim, Y.T.: A Study on Walking Stabilization and Path Tracking of Biped Robot Using RFID. Journal of Korean Institute of Intelligent Systems 23(1), 51–56 (2013)
8. Na, D.Y., Noh, S.H., Kim, Y.T.: Fuzzy Navigation and Obstacle Avoidance Control for Docking of Modular Robots. Journal of Korean Institute of Intelligent Systems 19(4), 470–477 (2009)
9. Jeong, D.H., Park, J.H., Park, J.I., Kim, Y.T.: A study on Omni-direction mobile robot for freight transportation. In: Conference on Information and Control System, pp. 137–138 (2012)

Design of Adaptive Network-Based Fuzzy Inference System for Obstacle Avoidance of Mobile Robot

Xi Li and Byung-Jae Choi

School of Electronic Engineering, Daegu University
Jillyang Gyeongsan-city Gyeonbuk, Korea
li_xi_lixi@hotmail.com, bjchoi@daegu.ac.kr

Abstract. There are many research results about obstacle avoidance in the field of intelligent mobile robots. Mobile robot can move in two-dimensional space from a set starting location to a known target location; in doing so, it must avoid any obstacles in its path as it seeks its goal. This paper uses the ANFIS method to record the path to the destination and to provide more rules to evaluate the situation. It is then possible to change the output and rules of the fuzzy inference system to obtain a better result. We show the feasibility of obstacle-avoidance performance using the ANFIS method.

Keywords: ANFIS, mobile robot, obstacle avoidance.

1 Introduction

Obstacle avoidance is a basic and important function of a mobile robot. Fuzzy inference system and neural network system are two very important and active branches in the field of intelligent control. These two control methods have significant superiority, but there are some limitations; therefore, a good combination of two methods is an inevitable result. Combination with a neural network enhances self-learning abilities and greatly simplifies data calculation. A fuzzy inference system is an effective tool for processing unknown models or difficult-to-establish model systems. However, the design of a fuzzy inference system doesn't depend on the object model, but is quite dependent on the experience and knowledge of experts or the operator; lack of such experience makes it difficult to obtain satisfactory control results.

A neural network has a strong self-learning ability, but generally does not deal directly with structured knowledge. It needs a large amount of training data, through the learning process and with its parallel distribution structure mapping relationship, to estimate the mapping relationship of input to output.

The adaptive network-based fuzzy inference system (ANFIS) is a neural network based on the Takagi–Sugeno fuzzy inference system. This paper is organized as follows: in the next section, we introduce the kinematics model of an indoor mobile robot. We describe the design process of the Sugeno-type fuzzy inference–based obstacle avoidance algorithm for an indoor mobile robot in Section 3. In Section 4, we

Y.-T. Kim, I. Kobayashi, and E. Kim (eds.), *Soft Computing in Advanced Robotics*,
Advances in Intelligent Systems and Computing 269,
DOI: 10.1007/978-3-319-05573-2_8, © Springer International Publishing Switzerland 2014

describe the design of an ANFIS system to record the path to the destination, which has more rules to evaluate the situation. We also change the output and rules of the fuzzy inference system to get a better result in this Section. Using Matlab/Simulink as a simulation tool, we show the feasibility of obstacle-avoidance using the ANFIS method in Section 5.

2 Architecture of Mobile Robot

The simulation view is shown in Figure 1 (a), where α is the angle of the mobile robot, β is the angle of the goal, and θ is the angle between the robot and the goal. The relationship among them is θ = α - β. The kinematics model of the robot is as shown in Figure 1 (b). Acceleration of the left and right wheels are ω_L and ω_R, respectively. ω is the angular velocity of the centered, L is the distance between two wheels, x and y are the position of the mobile robot. We assume that the contact between the wheels and the floor is pure rolling and non-slip [4].

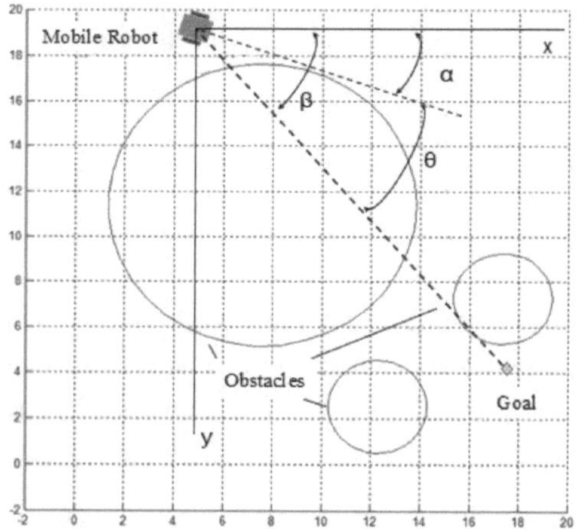

Fig. 1. (a)The simulation view

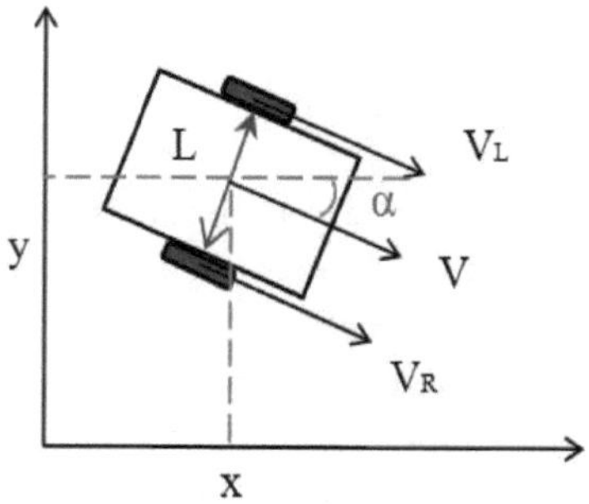

Fig. 2. (b) Kinematics model of the robot

The relationship between left and right wheels for velocity and acceleration is as follows:

$$V_L = r\omega_L \text{ and } V_R = r\omega_R \text{ ,} \tag{1}$$

where r is the radius of the wheel. The linear velocity of the mobile robot is V. Linear velocities for the left and right wheels are V_L and V_R , respectively. The relationship among ω_L, ω_R, V_L and $V_R$is as follows:

$$V = \frac{V_L + V_R}{2} \tag{2}$$

Now we can summarize a dynamic model for the mobile robot as follows:

$$\dot{x} = V \sin\alpha, \dot{y} = V\cos\alpha \ , \dot{\alpha} = \omega \ , \tag{3}$$

$$\begin{bmatrix} \dot{x} \\ \dot{y} \\ \dot{\alpha} \end{bmatrix} = \begin{bmatrix} \sin\alpha & 0 \\ \cos\alpha & 0 \\ 0 & 1 \end{bmatrix} \begin{bmatrix} V \\ \omega \end{bmatrix} \tag{4}$$

3 Design of Obstacle Avoidance Algorithm Based on Fuzzy Inference System

3.1 Obstacle Avoidance

There are some conventional obstacle-avoidance methods, such as edge-detection, certainty grids, and potential field methods. Ultrasonic obstacle avoidance uses the principle of measuring distance by ultrasound to detect obstacles around the mobile robot and then avoid them. Advantages to ultrasound obstacle avoidance are low cost and simple technological maturity; therefore, it is commonly used for obstacle avoidance. But there are also disadvantages, such as the direction of ultrasound is not good, precision is relatively low, and blind spot detection is unavailable.

3.2 Design of Fuzzy Inference System

Matlab code for the mobile robot's kinematics has already been written, so we added two constant blocks as the inputs to the mobile robot block. Constant blocks give the mobile robot two initial velocities for the left and right wheels.

When a fuzzy logic control system receives output data of the mobile robot block, it will generate an appropriate output according to fuzzy logic rules, and adds two initial velocities; therefore this system has the ability to change its path to reach the destination.

According to the control system model, we add three inputs into the fuzzy system: 1) theta, the measured angle between the mobile robot and the goal (θ); 2) distance, or the measured distance between the robot and the target; and 3) obstacle, or the measured obstacles detected.

The output of the fuzzy logic system is the changing speed values compared to previous velocities. There are three values (0.5, 0, and -0.5) that stand for the three types of speed changes during movement. The membership functions of input and output are shown in Figure 2. The rules of the fuzzy inference system are shown in Table 1.

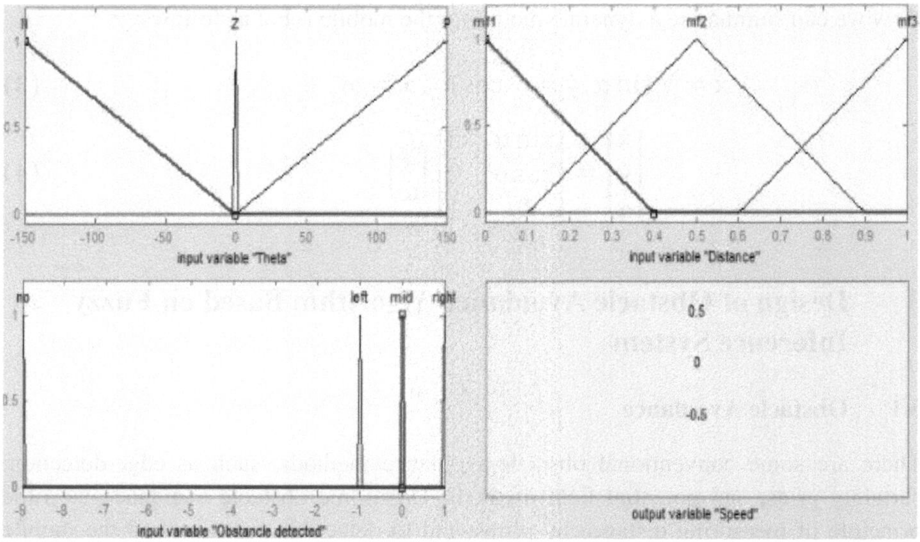

Fig. 2. The membership functions of inputs and output

Table 1. The rules for fuzzy inference system

Speed		*Obstacle detected*			
		left	mid	right	No
Theta	N	0.5	-0.5	-0.5	-0.5
	Z	0.5	-0.5	-0.5	0
	P	0.5	0.5	-0.5	0.5

4 Design of Adaptive Network-Based Fuzzy Inference System

The adaptive network-based fuzzy inference system (ANFIS) is the combination of fuzzy logic system and neural network system. Using a given input/output data set, the membership function parameters for fuzzy inference system (FIS) are tuned by using either a back-propagation algorithm alone or in combination with a least-squares method [6]. This adjustment allows fuzzy systems to learn from the data they are modeling. The modeling approach used by ANFIS is similar to many system identification techniques. Hypothesize a parameterized model structure. Collect input/output data in a form that will be usable by ANFIS for training. Then use ANFIS to train the FIS model to emulate the training data presented to it by modifying the membership function parameters according to a chosen error criterion.

4.1 Design of ANFIS

ANFIS information from Matlab is shown in Figure 3(a).In the new structure of the fuzzy logic system, we can see the rules change to 36 as shown in Figure 3(b). The new inputs and the surface of the fuzzy logic system are shown in Figures 4. Successful results from the Matlab simulation are shown in Figure 5.

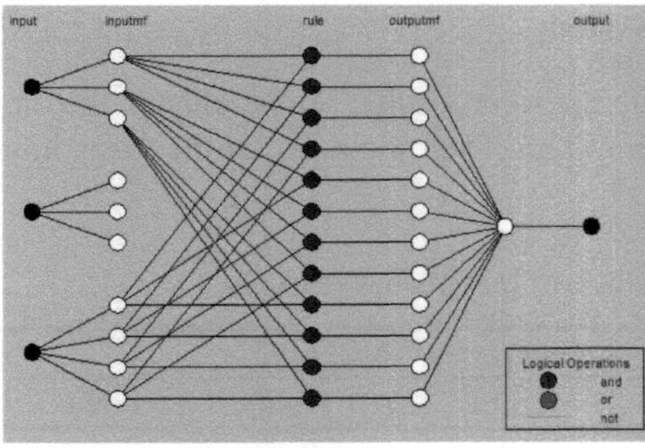

Fig. 3. (a) ANFIS

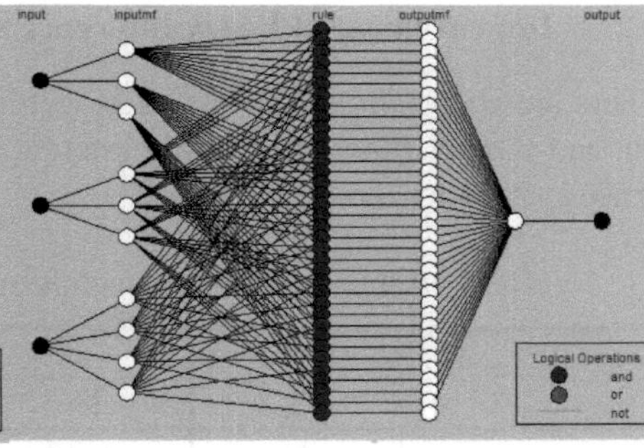

Fig. 3. (b) New structure of fuzzy logic system

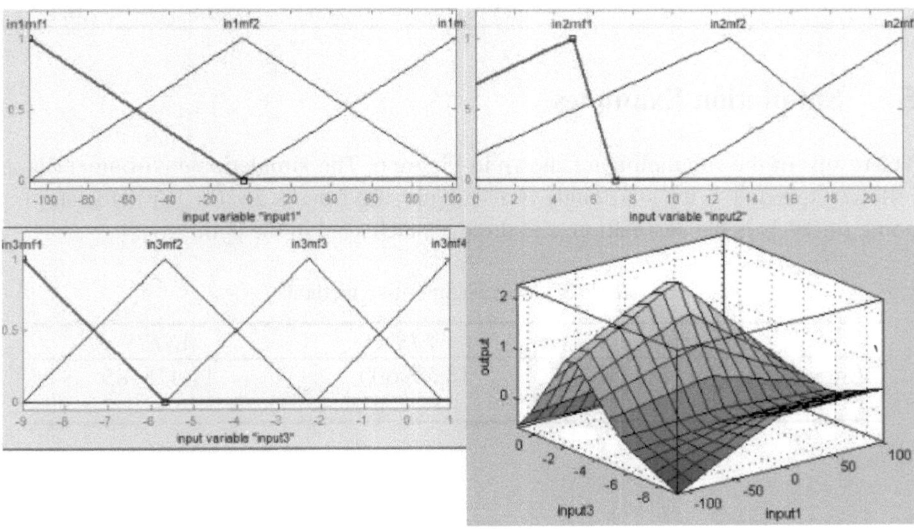

Fig. 4. The new inputs and the surface of the fuzzy logic system

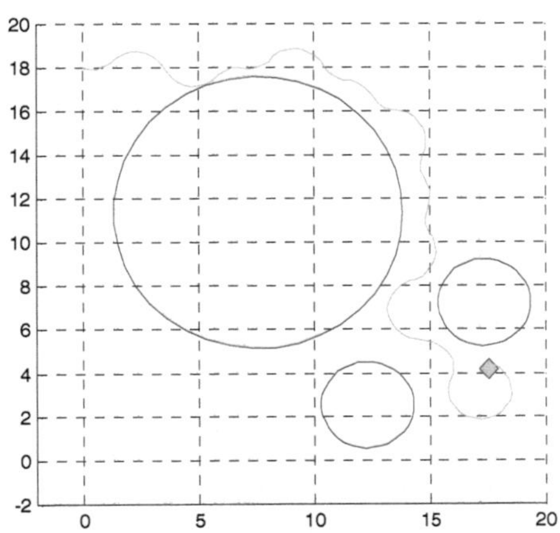

Fig. 5. Result of simulation

4.2 Improvement of Fuzzy Inference System

There are still many problems before getting the right results. So we change the output and control rules of the fuzzy inference system. Then the rules for the fuzzy inference system are shown in Table 2.

Table 2. Improvement of the rules for fuzzy inference system

Speed		*Obstacle detected*			
		left	mid	right	No
Theta	N	0.13	-0.5	-0.13	-0.5
	Z	0.5	-0.5	-0.5	0
	P	0.5	0.5	-0.5	0.5

5 Simulation Examples

The results of the simulation are shown in Figure 6. The simulation environment is as follows: speed is 5, the start angle is 1.983*pi, stop time is 56, (0, 18) is the starting point, (4, 17.5) is the destination, and three obstacles are in the path.

Table 3. Comparison of two methods

Method	*FIS*	*ANFIS*
Simulation Time (seconds)	13.599600	11.476185

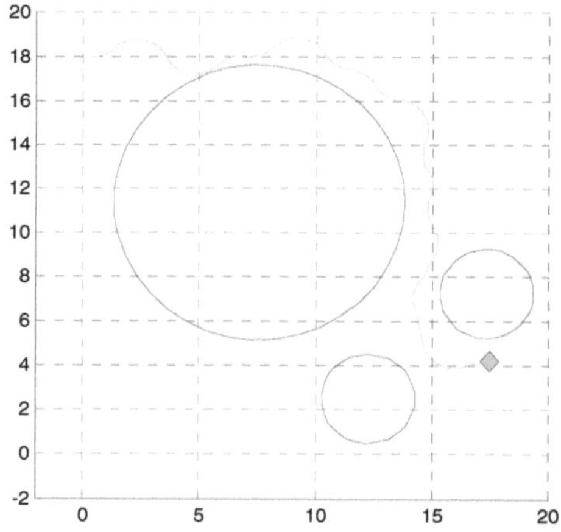

Fig. 6. The result of simulation

The comparison of the two methods is presented in Table 3; we can see that the ANFIS method is faster than the case of conventional FIS method.

We did some simulations in different environments, and we could also confirm that the robot moves to the goal to avoid obstacles successfully. The results are shown in Figure 7.

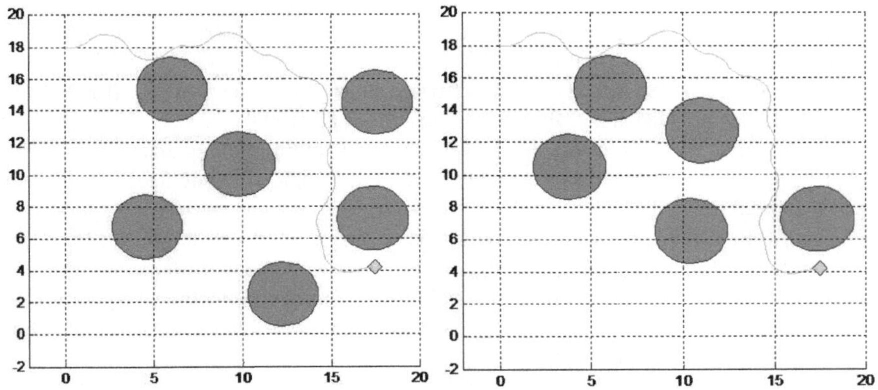

Fig. 7. Some simulation results

6 Concluding Remarks

An ANFIS overcomes the artificial decisions of fuzzification and defuzzification for FIS; fuzzy rules are not comprehensive and they are rough. Neural network technology was introduced, and it was more efficient to describe a human knowledge and to deal some situations with imprecise and uncertain situations. These strengthen the system against an unknown or changing environment in order to adjust the performance of the system itself. It does towards an adaptive, self-organizing, self-learning direction with the help of information storage ability and the learning ability of a neural network. In the fields of control, model prediction, pattern recognition, intelligent instruments, household appliances industry and other parts, ANFIS will be more widely used.

We used the ANFIS method to record the path to the destination. It has more rules to evaluate the situation. We changed the output and rules of the fuzzy inference system to get a better result. Through several simulations, we got that the ANFIS gives better results than the case of the conventional FIS.

References

1. Eustice, R.M., Singh, H., Leonard, J.J.: Exactly sparse delayed state filters for view based SLAM. IEEE Transaction on Robotics 22(6), 1110–1114 (2006)
2. Sala, P., Sim, R., Dickinson, S.: Landmark selection for vision-based navigation. IEEE Transaction on Robotics 22(2), 334–348 (2006)

3. Tong, W., Cheng, C., Hongche, G.: Research on obstacle-avoidance control algorithm of rehabilitation robot combined with neural network. In: 2010 Int. Conference on Measuring Technology and Mechatronics Automation, pp. 128–131 (2010)
4. Yang, S.X., Meng, M.: An efficient neural network approach to dynamic robot motion planning. Neural Networks 13(2), 143–148 (2000)
5. Wu, B., Shi, Z.: An ant colony algorithm based partition algorithm for TSP. Chinese Journal of Computers 24(12), 1328–1333 (2001)
6. Jang, J.-S.R.: ANFIS: Adaptive-Network-based Fuzzy Inference Systems. IEEE Transactions on Systems, Man, and Cybernetics 23(3), 665–685 (1993)
7. Tack, H.-H., Kim, C.-G., Kang, S.-C.: Tele-operated control of an autonomous mobile robot using a virtual force-reflection. Int. J. of Fuzzy Logic and Intelligent Systems 3(2), 244–250 (2003)
8. Jang, J.-S.R., Gulley, N.: Gain scheduling based fuzzy controller design. In: Proc. of the International Joint Conference of the North American Fuzzy Information Processing Society Biannual Conference, the Industrial Fuzzy Control and Intelligent Systems Conference, and the NASA Joint Technology Workshop on Neural Networks and Fuzzy Logic, San Antonio, Texas (December 1994)
9. Jin, T.: Obstacle avoidance of mobile robot based on behavior hierarchy by fuzzy logic. Int. J. of Fuzzy Logic and Intelligent Systems 12(3), 245–249 (2012)
10. Bae, Y., Kim, J., Kim, Y.: Obstacle avoidance methods in the chaotic mobile robot with integrated some chaos equation. Int. J. of Fuzzy Logic and Intelligent Systems 3(2), 206–214 (2003)
11. Kwon, K.-Y., Joh, J.: Autonomous navigation of an underwater robot in the presence of multiple moving obstacles. Int. J. of Fuzzy Logic and Intelligent Systems 5(2), 124–130 (2005)

Swarm EKF Localization for a Multiple Robot System with Range-Only Measurements

Shigekazu Fukui and Keitaro Naruse

University of Aizu
Tsuruga, Ikki-machi, Aizu-Wakamatsu, Fukushima 965-8580, Japan
{d8141102,naruse}@u-aizu.ac.jp

Abstract. Swarm localization, cooperative robot localization in swarm robotics, has a significant role in a swarm robot system and requires much deliberation for its estimation scheme. As such, designing stochastic hidden Markov model, in a way a variety of conditionally dependent, observed random variables such as measurements are effectively chosen and properly integrated into the probability distribution of a belief, is very important. In this paper, we propose swarm EKF localization, a hybrid of two inference algorithms, extended Kalman filter (EKF) and belief propagation (BP), with a capability of choosing how many dependencies of random variables are exploited in inference using the concept of neighborhood. Also, this paper presents a numerical experiment result of swarm EKF localizations. In conclusion, we could confirm that 2nd order neighborhood EKF has an overall better estimation performance compared to conventional 1st order neighborhood EKFs.

Keywords: extended Kalman filter, belief propagation, divide and conquer algorithm, swarm robotics, multi-agent systems.

1 Introduction

Swarm localization, a cooperative robot localization in swarm robotics, has a significant role in a swarm robot system where multiple robots operate autonomously, and requires much deliberation to design its technical estimation scheme. By nature, a swarm robot system should be a distributed one, yet major estimation techniques are mostly centralized one. To perform a good inference, the system should have a good view of the whole, but the system is not quite a whole, rather it is separated into each unit. Of course, a swarm robot system within a single, fairly centralized estimation scheme can be designed, or maybe standalone estimations performed by individual, fairly localized units of the system are enough for the estimation of the system.

However, the former and latter are extremes of what is centralized and what is distributed, respectively. In the former centralized case, as the number of robots in the system increases, the computational cost of inference, complexity of maintaining control over all robots, and message exchange between central and all robots will soon become out of control. Yet information available in inference, control and message exchange is abundant and whole, overall operation of the system will be

Y.-T. Kim, I. Kobayashi, and E. Kim (eds.), *Soft Computing in Advanced Robotics*,
Advances in Intelligent Systems and Computing 269,
DOI: 10.1007/978-3-319-05573-2_9, © Springer International Publishing Switzerland 2014

expected to be optimally accurate. In the latter distributed case, because inference performed by each robot becomes too localized to just a small and restricted amount of information, the result of inference will overfit to this insufficient information, and will converge to one of inferior estimates. Yet information reduction in inference, control and message exchange will benefit the system to reduce the operational load, overall operation of the system will be expected to be optimally economic.

With the advantage and disadvantage of two extremes are described and compared, one natural notion arises: *is there any intermediate of these two extremes?* A short answer will be *yes*; by virtue of divide and conquer (D&C) algorithm being a widely accepted, basic and efficient algorithm for all kinds of problems, there must be D&C-style formulation capable of choosing a point in a fine granularity between what is centralized and what is distributed for the localization problem of a swarm robot system. In this paper, as a hybrid of extended Kalman filter (EKF) and belief propagation (BP), we propose swarm EKF localization. This localization scheme is intended for a distributed swarm robot system with capability of range measurement and messaging among neighboring robots, and it has a relaxed distributed property compatible with a centralized one by exploiting interdependencies among robots using the concept of neighborhood.

Here is a list of the relevant literatures on EKF and BP. Kschischang et al. showed that Kalman filter, the original of EKF for a linear dynamic system, as a special case of BP (sum-product algorithm) [1]. Djugash et al. developed a distributed EKF with BP on a minimum spanning tree for a multiple robot system with range measurement [2]. In the field of material physics, Kikuchi proposed cluster variation method for defining a generalized locality for BP [3], and Yedidia et al. further generalized other BP methods such as junction graph method and cluster variation method, and developed region graph method [4].

2 Related Works

2.1 Extended Kalman Filter

Extended Kalman filter is a recursive state estimator for a nonlinear dynamic system [5]. It is statistical in outline, yet deterministic in detail; it uses Bayes filter and hidden Markov model (HMM) as its theoretical basis, and Gaussian distribution, linear transformation and minimization of mean square error as its practical interpretation and implementation.

EKF in general consists of two stages: prediction and measurement update. The prediction stage, which performs transformation of the belief distribution by control, is defined by the following equations:

$$\bar{\boldsymbol{\mu}}_t = \mathbf{g}\left(\mathbf{u}_t,\, \boldsymbol{\mu}_{t-1}\right) \ , \tag{1}$$

$$\bar{\boldsymbol{\Sigma}}_t = \mathbf{G}_t\, \boldsymbol{\Sigma}_{t-1}\, \mathbf{G}_t^T + \mathbf{R}_t \ . \tag{2}$$

Here, $\bar{\boldsymbol{\mu}}_t$ is the predicted belief vector, $\mathbf{g}$ is the state transition vector function, $\mathbf{u}_t$ is a control vector, $\boldsymbol{\mu}_{t-1}$ is the prior belief vector, $\bar{\boldsymbol{\Sigma}}_t$ is the predicted belief covariance matrix, $\mathbf{G}_t$ is the Jacobian matrix function of $\mathbf{g}$ with respect to the belief, evaluated with $\mathbf{u}_t$ and $\boldsymbol{\mu}_{t-1}$, $\boldsymbol{\Sigma}_{t-1}$ is the prior belief covariance matrix, and $\mathbf{R}_t$ is the state transition error covariance matrix model.

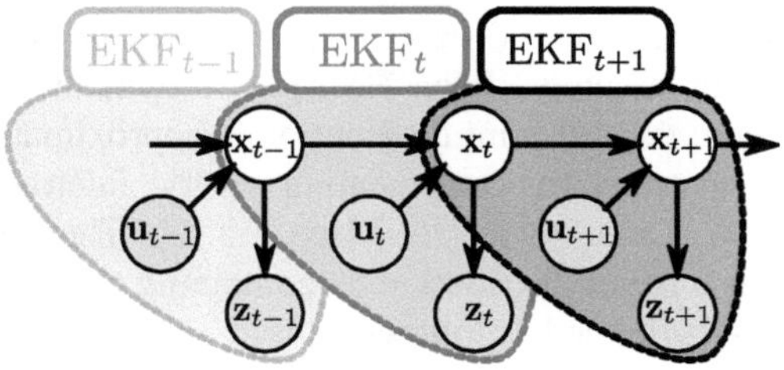

Fig. 1. HMM-style dynamic Bayesian network representing the flow of inference i.e. estimation performed by EKF. A shaded node is an observed variable; otherwise a node is a latent variable. The encircled area is the function window of EKF.

The measurement update stage, which performs innovation of the predicted belief by observation i.e. measurement, is defined by the following equations:

$$\mathbf{K}_t = \bar{\boldsymbol{\Sigma}}_t \mathbf{H}_t^T \left(\mathbf{H}_t \bar{\boldsymbol{\Sigma}}_t \mathbf{H}_t^T + \mathbf{Q}_t\right)^{-1} , \tag{3}$$

$$\boldsymbol{\mu}_t = \bar{\boldsymbol{\mu}}_t + \mathbf{K}_t \left(\mathbf{z}_t - \mathbf{h}\left(\bar{\boldsymbol{\mu}}_t\right)\right) , \tag{4}$$

$$\boldsymbol{\Sigma}_t = \left(\mathbf{I} - \mathbf{K}_t \mathbf{H}_t\right) \bar{\boldsymbol{\Sigma}}_t . \tag{5}$$

Here, $\mathbf{K}_t$ is the Kalman gain, $\mathbf{H}_t$ is the Jacobian matrix of $\mathbf{h}$ with respect to the belief, evaluated with $\bar{\boldsymbol{\mu}}_t$ (read further for $\mathbf{h}$), $\mathbf{Q}_t$ is the observation noise covariance model matrix, $\boldsymbol{\mu}_t$ is the updated, posterior belief vector, $\mathbf{z}_t$ is an observation, $\mathbf{h}$ is the observation model vector function, and $\boldsymbol{\Sigma}_t$ is the updated, posterior belief covariance matrix.

An illustration of HMM-style dynamic Bayesian network for the latent and observed random variables mentioned above and their conditional dependencies is given in Fig. 1. We could see EKF as an estimation function with a function window over localized sequential time frames and a portion of adjacently traceable dependencies among variables.

2.2 Belief Propagation

Belief propagation, first introduced by Judea Pearl, is a message passing, inference algorithm for graphical models such as Bayesian network and random Markov field (RMF) [1, 4, 6]. This algorithm is based on factorization of a joint distribution of various random variables with respect to probabilistic chain rule, and can be considered to be akin to divide and conquer algorithm; the process of multiplying factors one by one to form a product i.e. the exact or an approximated joint distribution of the original, can be considered as the process of acquiring related beliefs about each variable and accumulating them into the whole belief of the entire variables. Although, from a standpoint of factorization, this algorithm is concerned with inference of a joint distribution of many random variables in a given graphical model, calculating a marginal distribution with respect to variables of one's liking gives the desired distribution.

The process of BP has two distinct stages: message update and message composition. First, let us assume BP on HMM-style dynamic RMF for latent and observed variables where the full joint distribution is approximately factorized into a product of pairwise distributions representing every interdependencies of two variables e.g. $p(x_{t,j} \mid x_{t,i})$ and $p(\bar{x}_{t,j} \mid y_{t,i})$ (see Fig. 2). Then, the message update stage, which establishes a belief distribution by combining message distributions from neighborhood variables, is defined as follows:

$$p\left(x_{t,i} \mid y_{t,i}\right) = \eta\, p\left(\bar{x}_{t,i} \mid y_{t,i}\right) \prod_{j \in N^{\circ}(i)} m_{ji}\left(x_{t,i}\right) , \tag{6}$$

where $x_{t,i}$ is the posterior random variables of the i-th state after the message update, $\bar{x}_{t,i}$ is the prior random variable of the i-th state before the message update, $y_{t,i}$ is the observation of $x_{t,i}$, η is a normalization constant that, conforming to probability distribution in general, ensures the right-hand side of the equation to be summed to 1, $N^{\circ}(i)$ is the open neighborhood set of the i-th state (*open* means i is not included in a set; *closed* means vice versa), and $m_{ji}(x_{t,i})$ is a message from $x_{t,j}$ to $x_{t,i}$.

Remark 1. In Bayesian probability, especially in a statistical model with incomplete random variables (*complete* or *incomplete*, with a sense that the statistics on values of a random variable, i.e. the probability distribution of the variable, does or does not convey the exact information of what value the variable will take in a given situation where the statistics were compiled), $m_{ji}(x_{t,i})$ is equivalent to $p(x_{t,i})$, yet does not uniquely and exactly represent $p(x_{t,i})$. To readers, think each $m_{ji}(x_{t,i})$ as an instance of $p(x_{t,i})$ with varying completeness, and multiplying them leads to a higher completeness of $p(x_{t,i})$. The resultant distribution of Eq. (6) has a reasonably weak probability at a specific value of $x_{t,i}$ where each $m_{ji}(x_{t,i})$ has a particularly weak probability, and the statement vice versa also holds for a strong probability of $p(x_{t,i})$.

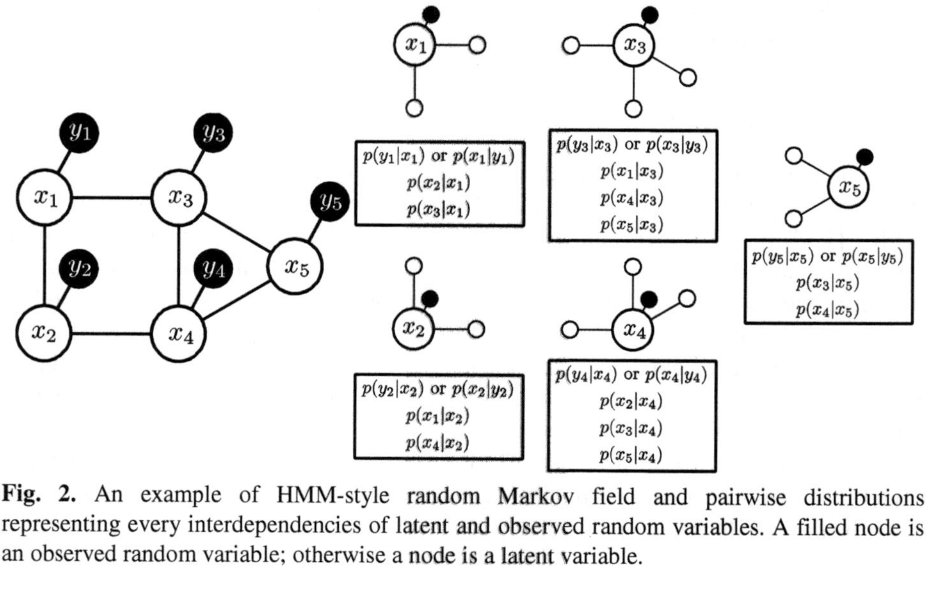

Fig. 2. An example of HMM-style random Markov field and pairwise distributions representing every interdependencies of latent and observed random variables. A filled node is an observed random variable; otherwise a node is a latent variable.

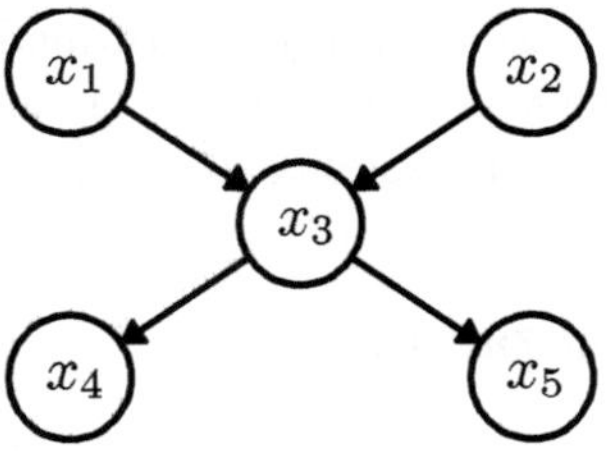

Fig. 3. An example of polytree Bayesian network

The following equation defines the message composition stage:

$$m_{ij}\left(x_{t,j}\right) = \eta \sum_{x_{i,t}} p\left(x_{t,j} \mid x_{t,i}\right) \left(p\left(x_{t,i} \mid y_{t,i}\right) \prod_{k \in N^{\circ}(i) \setminus \{j\}} m_{ki}\left(x_{t,i}\right) \right) , \tag{7}$$

where $p(x_{t,j} \mid x_{t,i})$ is a pairwise distribution that conceptually works as a messaging route from $x_{t,j}$ to $x_{t,i}$.

In this definition of BP on HMM-style dynamic RMF, knowledge of distributions such as pairwise distributions $p(x_{t,j} \mid x_{t,i})$ and $p(x_{t,j} \mid y_{t,i})$ is required. In a much generalized form of definition, one can factorize a full joint distribution into not just pairwise distributions but also distributions of one, three, four or more variables to define a different BP if there is always at least one messaging route between any two variables, and part of, not necessarily all of, dependencies of variables are used as messaging routes. With such factorization performed, required knowledge for distributions changes. For example, in BP on polytree Bayesian network shown in Fig. 3, the full joint distribution can be factorized into distributions without loosing dependencies of random variables, and if these factor distributions are exact, all marginal distributions are also exact:

$$p(x_1, x_2, x_3, x_4, x_5) = p(x_1, x_3|x_2)\, p(x_2)\, p(x_4|x_3)\, p(x_5|x_3) \tag{8}$$

$$\begin{aligned} &\sum_{x1,x2,x4,x5} p(x_1, x_2, x_3, x_4, x_5) \\ &\qquad = \sum_{x1,x2,x4,x5} p(x_1, x_3|x_2)\, p(x_2)\, p(x_4|x_3)\, p(x_5|x_3) \\ &\qquad = \sum_{x1,x2} p(x_1, x_2, x_3) \sum_{x4} p(x_4|x_3) \sum_{x5} p(x_5|x_3) \\ &\qquad = p(x_3) \ . \end{aligned} \tag{9}$$

Here, Eq. (8) is one example of factorization where dependencies of random variable are entirely preserved in the resultant product of distributions. In Eq. (9), with knowledge of probability distributions $p(x_1, x_3|x_2)$, $p(x_2)$, $p(x_4|x_3)$ and $p(x_5|x_3)$, the exact marginal distribution $p(x_3)$ is calculated. Next, in BP on acyclic RMF, contrary to the last example, the full joint distribution cannot be factorized into distributions without loosing entirety of interdependencies of random variables, and thus, despite the proven fact of BP that it does converge to a certain distribution in polytree graphical model, the exact marginal distributions cannot be obtained. At formulation of BP for a certain graphical model, a careful consideration should be taken concerning exactness of factorization and obtainability of required distributions.

3 Swarm EKF Localization

Our proposed swarm EKF localization, a hybrid of EKF and BP for a swarm robot system compatible to be both distributed and centralized, is covered in this section. For EKF and BP in general were discussed in the previous section, now the formulation of a robot swarm and actual adaptation of EKF and BP for our research will be describe in the following subsections.

3.1 Formulation of a Robot Swarm

In this swarm robot system, each robot is assumed to be equipped with a range sensor and a wireless communication device, and other robots within the range of both the sensor and communication device of the robot are considered as the neighborhood. By assumption, the range sensor obtains not only a range measurement but also the identity of the measured robot. Now, let us clarify and generalize definition of the neighborhood of a robot mentioned above. As already briefly mentioned in Section 2.2, employing a similar notion in graph theory, two types of neighborhood can be defined: a closed neighborhood $N^{\bullet}$ and an open neighborhood N°. And in this paper, notation of $N_k^{\circ}(i)$ means the k-th order open neighborhood of the i-th robot. The mathematical definition of closed and open neighborhood is as follows:

$$N_0^{\bullet}(i) = \{i\}, \tag{10}$$

$$N_1^{\bullet}(i) = \{i\} \cup \{a \,|\, \text{within\_range}(i,a) \wedge i \neq a\}, \tag{11}$$

$$N_k^{\bullet}(i) = N_{k-1}^{\bullet}(i) \cup \bigcup_{j \in N_{k-1}^{\bullet}(i) \setminus N_{k-2}^{\bullet}(i)} \{a \,|\, \text{within\_range}(j,a) \wedge j \neq a\}, \tag{12}$$

$$N_k^{\circ}(i) = N_k^{\bullet}(i) \setminus \{i\}, \tag{13}$$

where i, j and a are indices of certain robots in the system, and within_range (i,a) is a logical proposition being true if the a-th robot is within the range of the sensor and communication device of the i-th robot. Within a neighborhood, data such as messages (marginal belief distributions) and range measurements are exchanged to facilitate inference. An illustration of the 1st and 2nd order closed neighborhood for rectangular and triangular lattices is provided in Fig. 4. The swarm EKF localization scheme is characterized by its distributed local EKF instances, and the state—in this paper, location—of a robot are inferred in each distributed local EKF instances while gathering data from its k-th order neighborhood robots.

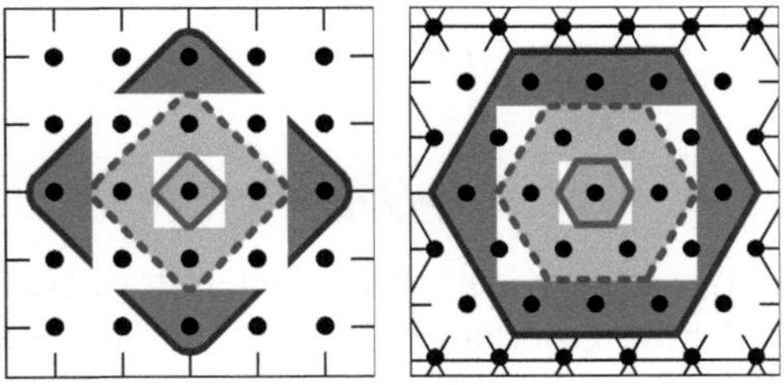

Fig. 4. An illustration of the 1st and 2nd order closed neighborhood for a rectangular lattice and a triangular lattice. A set of nodes inside the red dashed line is the 1st order closed neighborhood, and inside the outer solid blue line is the 2nd order closed neighborhood.

3.2 Adaptation of Extended Kalman Filter

State and Measurement Models. For analytical purpose, we assumed a stationary swarm robot system for the time being, so that the prediction stage defined by Eqs. (1) and (2) are left undefined and thus omitted here.

Now, let us assume a swarm robot system on the Cartesian space $\mathbb{R}^2$. The true position of the i-th robot is a vector $\mathbf{x}_{t,i} = (x_{t,i}\ ,\ y_{t,i})$, and corresponding to $\mathbf{h}$ in Eq. (4), the range measurement vector model from the i-th robot to its neighboring j-th robot is defined below:

$$\mathbf{h}_{r;t,i,j} = \left(\|\boldsymbol{\mu}_{t,i,\{i\}} - \boldsymbol{\mu}_{t,i,\{j\}}\|_2 \right) , \tag{14}$$

where j follows $j \in N_k^\circ(i)$, $\|\cdot\|_2$ is the $\mathbb{R}^2$ Cartesian norm, $\boldsymbol{\mu}_{t,i,\{i\}}$ and $\boldsymbol{\mu}_{t,i,\{j\}}$ are the positional beliefs of the i-th and j-th robot estimated by the i-th robot respectively (notion of $\{i\}$ in the subscript of variables such as $\boldsymbol{\mu}_{t,i,\{i\}}$ is explained at the end of this subsection). The reason of using the j-th robot belief estimated by the i-th robot instead of one estimated by the j-th robot is for multiplicity and redundancy of the j-th robot belief, which is expected to reduce the message exchange load and cancel the effect of localized measurement-estimation outliers. Similarly, the actual range measurement vector obtained through sensing corresponding $\mathbf{z}_t$ in Eq. (4) is:

$$\mathbf{z}_{r;t,i,j} = \left(\|\mathbf{x}_{t,i} - \mathbf{x}_{t,j}\|_2 \right) + \boldsymbol{\epsilon}_{r;t} , \tag{15}$$

and the range measurement noise covariance model matrix corresponding to $\mathbf{Q}_t$ in Eq. (3) is assumed to be:

$$\mathbf{Q}_{r;t} = \mathrm{cov}\left(\boldsymbol{\epsilon}_{r;t}\ ,\ \boldsymbol{\epsilon}_{r;t}\right) = \left(\sigma_{r;t}^2\right) , \tag{16}$$

where $\boldsymbol{\epsilon}_{r;t}$ is the range measurement noise vector that is assumed to be white noise, and $\sigma_{r;t}$ is the range measurement standard deviation model for $\boldsymbol{\epsilon}_{r;t}$.

Now, considering multiplicity and redundancy of the belief and assuming $N_k^\bullet(i)$ to be an ordered set, the whole belief estimated by the i-th robot corresponding to $\bar{\boldsymbol{\mu}}$ and $\boldsymbol{\mu}$ in EKF is defined as follows:

$$[s, e] \subset \mathbb{Z} \equiv N_k^\bullet(i) , \tag{17}$$

$$\boldsymbol{\mu}_{t,i} = \begin{pmatrix} \boldsymbol{\mu}_{t,i,\{s\}} \\ \vdots \\ \boldsymbol{\mu}_{t,i,\{e\}} \end{pmatrix}. \tag{18}$$

Here, $[s, e]$ is a bijective integer interval convenient for iteration over elements of $N_k^\bullet(i)$ in the specific order from s through e, and $\boldsymbol{\mu}_{t,i}$ is the whole belief block vector estimated by the i-th robot concerning the neighborhood $N_k^\bullet(i)$, consisting of vectors from $\boldsymbol{\mu}_{t,i,\{s\}}$ through $\boldsymbol{\mu}_{t,i,\{e\}}$; its dimension is $2\,|N_k^\bullet(i)|$. In the same manner, the whole belief covariance matrix estimated by the i-th robot corresponding to $\bar{\boldsymbol{\Sigma}}$ and $\boldsymbol{\Sigma}$ is defined as follows:

$$\boldsymbol{\Sigma}_{t,i} = \begin{pmatrix} \boldsymbol{\Sigma}_{t,i,\{s\}\{s\}} & \cdots & \boldsymbol{\Sigma}_{t,i,\{s\}\{e\}} \\ \vdots & \ddots & \vdots \\ \boldsymbol{\Sigma}_{t,i,\{e\}\{s\}} & \cdots & \boldsymbol{\Sigma}_{t,i,\{e\}\{e\}} \end{pmatrix}. \tag{19}$$

Here, $\boldsymbol{\Sigma}_{t,i}$ is the whole belief covariance block matrix estimated by the i-th robot concerning the neighborhood $N_k^\bullet(i)$, consisting of matrices $\boldsymbol{\Sigma}_{t,i,\{\cdot\}\{\cdot\}}$; its row and column dimensions are $2\,|N_k^\bullet(i)|$.

Lastly, defining $\mathbf{b}_{t,i}$ as the belief random vector variable of the i-th robot, the Jacobian matrix of $\mathbf{h}_{r;t,i,j}$, corresponding to $\mathbf{H}_t$ in the measurement update stage in EKF, with respect to the belief $\mathbf{b}_{t,i}$, evaluated with $\boldsymbol{\mu}_{t,i}$ is defined as:

$$\mathbf{H}_{r;t,i} = \left.\frac{\partial\,\mathbf{h}_{r;t,i,j}}{\partial\,\mathbf{b}_{t,i}}\right|_{\mathbf{b}_{t,i}=\boldsymbol{\mu}_{t,i}}. \tag{20}$$

The notion used throughout this subsection, $\{\cdot\}$ in the subscript of a variable or value, is an ordered set of robot indices of which this variable or value holds information. For example, for the whole belief vector, belief covariance matrix and belief random vector variable, $\boldsymbol{\mu}_{t,i} = \boldsymbol{\mu}_{t,i,[s,e]}$, $\boldsymbol{\Sigma}_{t,i} = \boldsymbol{\Sigma}_{t,i,[s,e][s,e]}$ and $\mathbf{b}_{t,i} = \mathbf{b}_{t,i,[s,e]}$.

3.3 Adaptation of Belief Propagation

Message Update and Message Composition. Simply put, both the message update and message composition stages can be implemented by the measurement update stage in EKF; dismissing multiplication of a belief propagation and a product of message propagations as shown in Eqs. (6) and (7), inference rather relies on minimization of the mean square error between each message and the message model evaluated with a belief. First, let us assume all probability distributions in our BP are normal distributions, and a multivariate probability distribution is replaced by a multivariate normal distribution conforming to the following form:

$$p(\boldsymbol{\alpha}\,|\,\boldsymbol{\beta}) \sim \mathcal{N}(\boldsymbol{\alpha};\,\boldsymbol{\mu}_{\boldsymbol{\alpha}}, \boldsymbol{\Sigma}_{\boldsymbol{\alpha}}\,|\,\boldsymbol{\beta}), \tag{21}$$

where $\boldsymbol{\alpha}$ is the random vector variable represented by this probability distribution, $\boldsymbol{\beta}$ is a random vector variable conditional to $\boldsymbol{\alpha}$, $\boldsymbol{\mu}_{\alpha}$ is the mean value of $\boldsymbol{\alpha}$, and $\boldsymbol{\Sigma}_{\alpha}$ is the covariance matrix of $\boldsymbol{\alpha}$. Taking account that messages are also probability distributions as mentioned in Remark 1 in Section 2.2, a message from the j-th robot to i-th robot is:

$$m_{ji}(x_{t,i}) = \mathcal{N}\left(\mathbf{b}_{t,j,\{i\}}; \boldsymbol{\mu}_{t,j,\{i\}}, \boldsymbol{\Sigma}_{t,j,\{i\}\{i\}}\right) . \tag{22}$$

Here, $\mathbf{b}_{t,i,\{i\}}$ is the belief random vector variable of the i-th robot maintained by the j-th robot. Similarly, corresponding to $p\left(\bar{x}_{t,i} \mid y_{t,i}\right)$and $p\left(x_{t,i} \mid y_{t,i}\right)$in Eq. (6), the prior and posterior belief distributions before and after the message update, estimated by the i-th robot are defined as follows:

$$\mathcal{N}\left(\bar{\mathbf{b}}_{t,i}; \boldsymbol{\mu}_{t,i}, \boldsymbol{\Sigma}_{t,i} | \mathbf{Z}_{t,i}\right) , \tag{23}$$

$$\mathcal{N}\left(\mathbf{b}_{t,i}; \boldsymbol{\mu}_{t,i}, \boldsymbol{\Sigma}_{t,i} | \mathbf{Z}_{t,i}\right) . \tag{24}$$

Finally, corresponding to $\mathbf{h}$ and $\mathbf{H}_t$ in the measurement update stage in EKF, the definition of the message vector model from the j-th robot to the i-th robot maintained by the i-th robot, and the Jacobian of the model with respect to $\mathbf{b}_{t,i}$ evaluated at $\boldsymbol{\mu}_{t,i}$ is:

$$\mathbf{h}_{m;t,i,ji}\left(\boldsymbol{\mu}_{t,i}\right) = \left(\boldsymbol{\mu}_{t,i,\{i\}}\right) , \tag{25}$$

$$\mathbf{H}_{m;t,i} = \left.\frac{\partial\,\mathbf{h}_{m;t,i,ji}}{\partial\,\mathbf{b}_{t,i}}\right|_{\mathbf{b}_{t,i}=\boldsymbol{\mu}_{t,i}} . \tag{26}$$

Here, a message from the j-th robot to the i-th robot is modeled as the i-th robot belief estimated by the i-th robot.

By substituting Eqs. (25) and (26) into the measurement update stage of EKF, the message update stage is formulated. For the message composition stage, the process is quite similar to the message update stage except a message received from the robot to whom the composed message is sent is not used, and the resultant posterior belief is not used for updating the self belief, rather it is marginalized and sent to the intended recipient robot as a message.

4 Numerical Experiments

In this section, the performance of the proposed swarm EKF localization scheme along with two other conventional EKF localization schemes is evaluated with numerical experiments. All experiments are performed under the same arrangement, a stationary 25-robot swarm, except that uncertainties of initial positional beliefs and range measurements are intentionally varied for accentuating the influence of input uncertainties to localization accuracy. Figure 5 provides an illustration of the robot swarm and its random Markov field, and one example from various predefined sets of initial localization beliefs. The RMF in Fig. 5 shows what neighborhood relations this robot swarm holds; as described in Section 3.1, this RMF provides a hint of where range sensing and massage exchange happen.

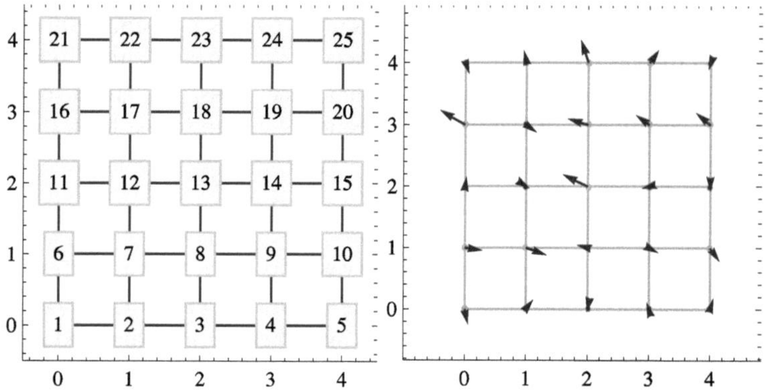

Fig. 5. An illustration of the stationary 25-robot swarm and its random Markov field used in experiments (left) and one example (standard deviation $\sigma = 0.2$) from predefined sets of the corresponding 25 initial localization beliefs (right). An arrow represents displacement of a belief from its true position.

The two conventional EKF localization schemes are instances of 1st order neighborhood EKF, thus they facilitate information localized to robots in 1st order closed neighborhood. The proposed swarm EKF localization scheme, for this experiment settings, is an instance of 2nd order neighborhood EKF, thus it facilitates information localized to robots in 2nd order closed neighborhood. Also, one of the 1st order neighborhood EKF schemes uses a minimum spanning tree (MST) of the network, which is characterized by absence of loops, as its messaging routes of BP, and the rest use the original loopy RMF as its messaging routes. Figure 6 provides an illustration of the minimum spanning tree and original graph used by messaging routes of BP.

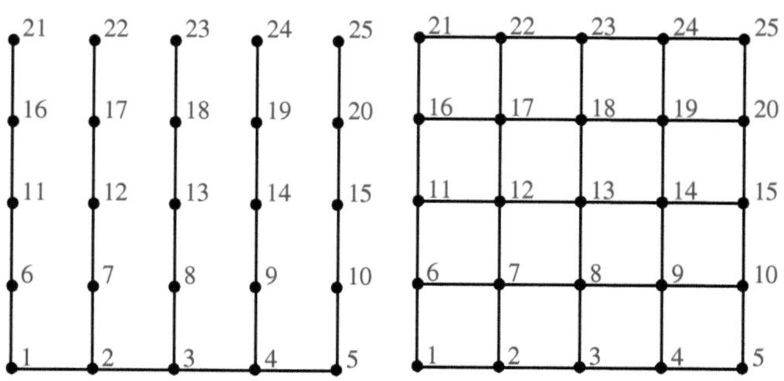

Fig. 6. An illustration of the minimum spanning tree (left) and the original graph (right) used by messaging routes of belief propagation

In both the MST and original graph, robot 1 is designated as the root node, and for the original loopy graph robot 25 is designated as the message starting node. Message exchange is performed sequentially and synchronously, so that starting from leaf nodes or specifically designated message starting nodes messages are composed and sent one by one at each node to the root node. Once the root node is reached, message exchange is performed again with the reverse order. And if message exchange reached back to all the starting nodes, then the message update stage is performed at every robot.

Table 1. Performance evaluation of each EKF scheme. The initial beliefs are at time t=1, and the final beliefs are at time t=100. *Small* and *large* uncertainties means the initial beliefs and measurements has uncertainty with standard deviation $\sigma = 0.2$ and $\sigma = 1.6$ respectively.

Extended Kalman filter	k-th order neighborhood	Messaging graph	Uncertainty	RMS error (RMS, root mean square): Initial beliefs	RMS error (RMS, root mean square): Final beliefs	Calc. time [s]	Belief vector size: Avg	Belief vector size: Min	Belief vector size: Max
Conventional method	1	MST	Small	0.2538	0.1858	13.67	8.4	6	10
	1	MST	Large	2.2606	2.9141	13.61	8.4	6	10
	1	Original graph	Small	0.2538	0.1765	19.92	8.4	6	10
	1	Original graph	Large	2.2606	2.0773	20.03	8.4	6	10
Proposed method	2	Original graph	Small	0.2538	0.0704	291.91	18.32	12	26
	2	Original graph	Large	2.2606	1.3180	289.84	18.32	12	26

RMS root mean square, *MST* minimum spanning tree.

Result. Statistics on localization error, calculation speed and dimension of a belief compiled from the two conventional EKF localization and the proposed swarm EKF localization schemes is shown in Tables 1 and 2. From root mean square (RMS) errors and error histograms, although in a large uncertainty all EKF schemes failed to localize their position accurately, regardless of the degree of uncertainties the proposed, exemplary 2nd order neighborhood swarm EKF localization scheme performed better than the other two conventional 1st order neighborhood EKF schemes. It should be contributed from a wider covering of dependencies among random variables, such that localized outliers are canceled by messages and range measurements gathered from reasonably many neighboring robots. One notable drawback of the proposal is the calculation time, which could be mainly caused by two factors: multiplicity and redundancy in representation of beliefs (see statistics on the dimension of beliefs) and sequentiality of the process of BP.

Table 2. An illustration of final beliefs (t=100) and its error histogram. *Small* and *large* uncertainties means the initial beliefs and measurements has uncertainty with standard deviation $\sigma = 0.2$ and $\sigma = 1.6$ respectively.

Extended Kalman filter	*k*-th order neighborhood	Messaging graph	Uncertainty	Final beliefs	Error histogram of final beliefs (RMS, root square mean)
Conventional method	1	MST	Small	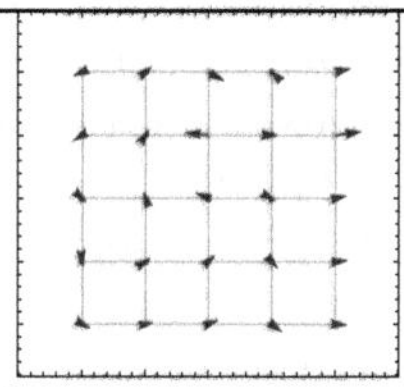	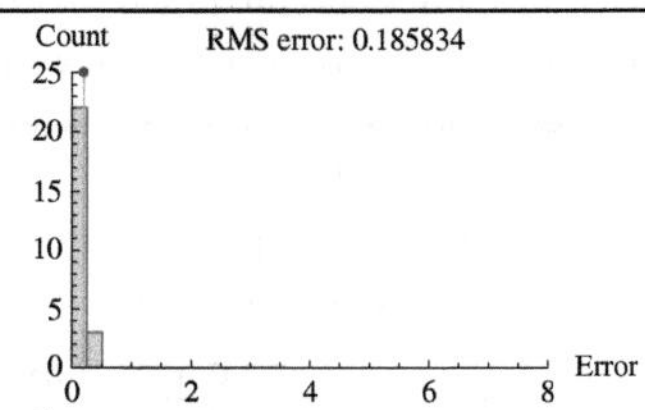
	1	MST	Large	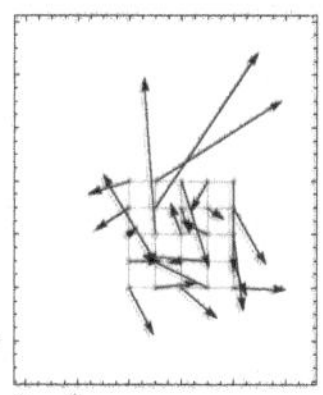	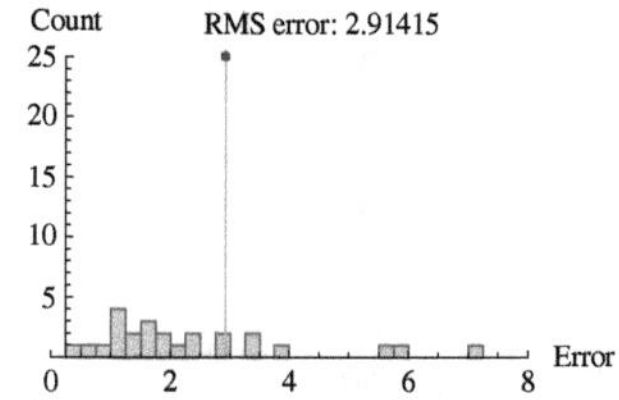
	1	Original graph	Small	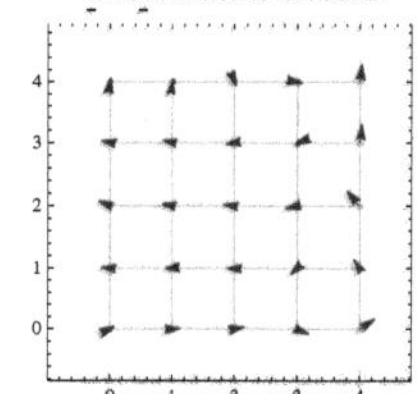	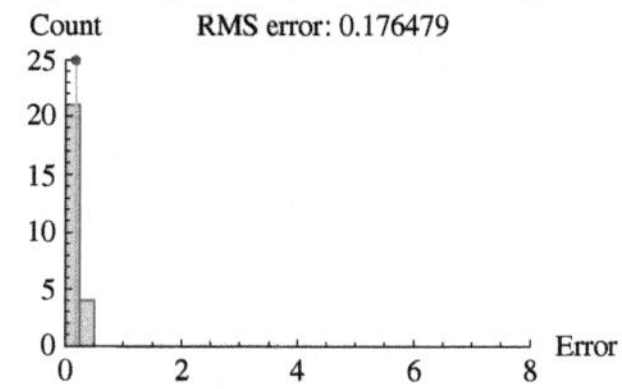
	1	Original graph	Large	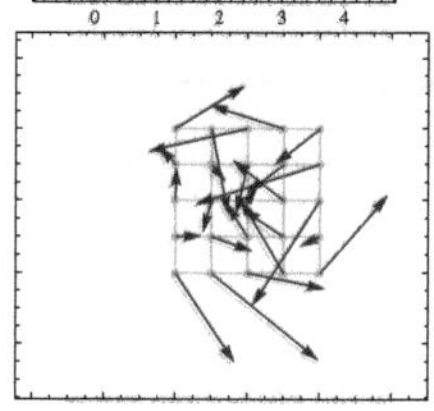	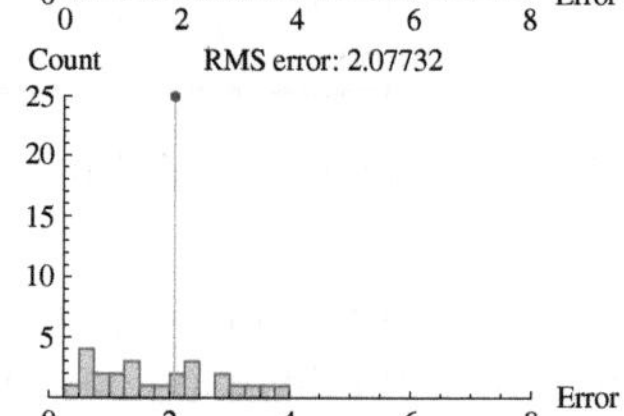
Proposed method	2	Original graph	Small	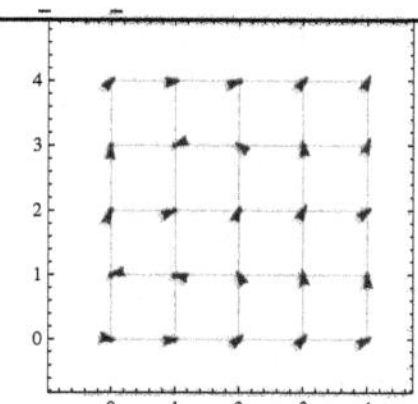	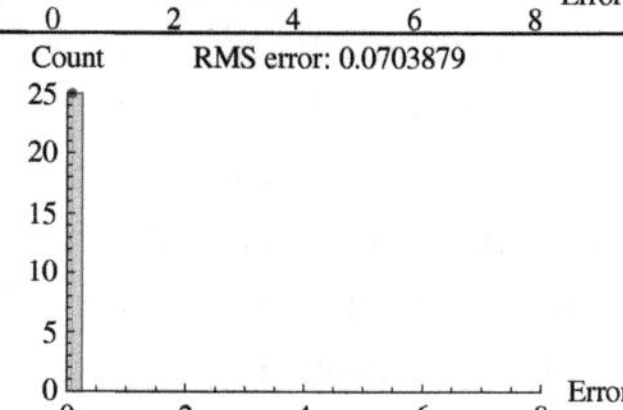
	2	Original graph	Large	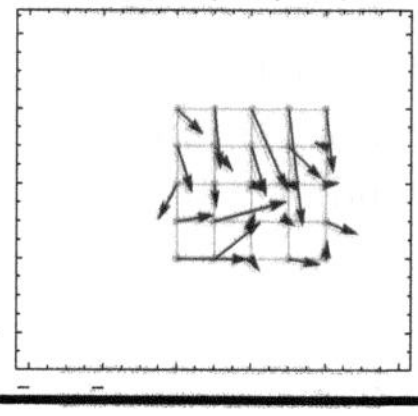	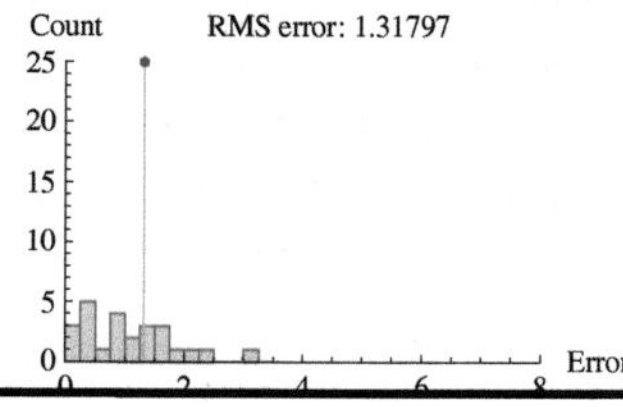

MST minimum spanning tree.

5 Conclusion

Following the paradigm of divide and conquer (D&C) algorithm, we closely examined extended Kalman filter (EKF) and belief propagation (BP) in search of an intermediate formulation that fits between centralization and distribution of a swarm robot system, and combined them together to devise an EKF-BP hybrid localization scheme compatible to be both centralized and distributed, called swarm EKF localization. The performance of the exemplary 2nd order neighborhood scheme of the proposal, in contrast to two conventional 1st order neighborhood schemes, which by themselves can fall within the varying "neighborhood scheme" of the proposal, was confirmed to be better in localization accuracy. As for the drawback of long calculation time, the same asynchronous BP formulation reported by Djugash [2], which has reasonable accuracy while greatly lessen the calculation time, can be used.

Although methodology of dividing the swarm robot localization problem into problem subunits and integrating/accumulating the solutions of these subunits was investigated from the angle of BP, it was only within a paradigm of neighborhood characterized by spatial range. If a swarm localization problem is expanded to a swarm control problem that pursues the system to be in certain goal state in a given state transition model i.e. a probabilistic automaton, such that a particular topology of the robot swarm must be maintained or a particular location must be reached by at least one robot, the resultant graphical model representing the problem will become larger and more complex, yet will leave a plenty of opportunities to be conquered by D&C algorithm, thus also by our swarm EKF scheme.

References

1. Kschischang, F.R., Frey, B.J., Loeliger, H.A.: Factor graphs and the sum-product algorithm. IEEE Trans. Inf. Theory 47(2), 498–519 (2001)
2. Djugash, J., Singh, S.: Motion-aided network SLAM with range. IJRR 31(5), 604–625 (2012)
3. Pelizzola, A.: Cluster Variation Method in Statistical Physics and Probabilistic Graphical Models. J. Phys. A: Math. Gen. 38(33) (2005)
4. Yedidia, J.S., Freeman, W.T., Weiss, Y.: Understanding Belief Propagation and its Generalizations. In: Lakemeyer, G., Nebel, B. (eds.) Intelligence, Citeseer, vol. 8(1), pp. 236–239 (2002)
5. Thrun, S., Burgard, W., Fox, D.: Probabilistic Robotics (Intelligent Robotics and Autonomous Agents). The MIT Press (2005)
6. Pearl, J.: Probabilistic Reasoning in Intelligent Systems: Networks of Plausible Inference. Morgan Kaufmann (1988)

Inspection Robot for Parallel Entry Boiler Header Pipe

Adzly Anuar*, Nur Shahida Roslin,
Khairul Salleh Mohamed Sahari, and M. Azwan Aziz

Center for Mechatronics and Advanced Robotics, Universiti Tenaga Nasional, Malaysia
adzly@uniten.edu.my

Abstract. This paper presents the development of inspection robot for parallel entry boiler header pipes with drastic change in diameter. The proposed robot consists of caterpillar based robot body, lifting mechanism and camera gimbal mechanism. These mechanisms are designed to adapt the sudden pipe diameter change from 89mm to 300mm. This robot applies differential drive locomotion system that provides turning ability plus forward and reverse movement. Lifting mechanism consists of linear transmission using leadscrew and scissors linkage mechanism. This enables the robot to retract when entering the parallel entry and expand in bigger pipe. Camera gimbal mechanism enables the robot to acquire 360° inner pipe surface image with its position. Experiments have been accomplished to investigate the functionality of these mechanisms. Results obtained shows that the robot has fulfilled its design requirements and achieved its objective.

1 Introduction

Pipelines are the major components in power plant industry, oil and gas industry, food industry and sewage industry. Inspections are demanded to ensure the pipelines are in the good condition and to avoid explosion and poisonous food supplies. However, most pipelines in the industry are very lengthy and hazardous to human being as the condition in the pipelines is hot, lack of oxygen supply and might contain chemical substances. This is the reason why pipe inspection robots are one of the growing technologies in the world. Research in this area has been done for the past two decades. Boiler header is one of the crucial pipelines thermal power plants and usually requires periodical maintenance and inspection. Routine maintenance check allowed only small opening which is either by cutting the nipple part or handheld nipple. The nipple size is in range between 89 mm and 102 mm relatively small as compared to the boiler header which is in range between 200mm and 300 mm [1].

Many designs of robots have been developed to accomplish their own requirements. A pipe inspection robot using vibration energy for camera actuator has been developed for 50mm diameter pipe [2]. Modular type robot for pipe diameter range 22mm and 35mm can be manually reconfigured to fulfill certain application at a time [3]. A snake type robot using helical rolling motion with conical spiral form is de-

* Corresponding author.

Y.-T. Kim, I. Kobayashi, and E. Kim (eds.), *Soft Computing in Advanced Robotics*,
Advances in Intelligent Systems and Computing 269,
DOI: 10.1007/978-3-319-05573-2_10,

signed for locomotion system and better image capturing ability [4]. Another snake type robot using sinusoidal wave drive has been tested to cater small diameter change which is 55mm to 80mm but no result on image acquisition [5]. Wheel wall press type using screw drive locomotion system equipped with two cameras for navigation and detection ability has been invented for 175mm to 205mm diameter pipes [6]. A pipe inspection gauge (PIG) type robot is built for 13mm pipelines using compressed air blowing and wire guided as its locomotion and has a small camera module for inspecting purposes [7]. Caterpillar type of robot, mostly to encounter a rough environment, also known as crawler has been invented as modular to be assembled with three other crawlers acting as the wheel and flippers to a robot body [8] [9]. Triangular shape crawler robot has been proposed for high-step climbing purposes [10]. However, none of this invention has been proved to perform sudden change diameter inspection activity in 89mm to 300mm pipe.

2 Design Requirement and Specification

Design requirements are obtained based on the raw data collected from a local power plant. As the result, following requirements are established:

- Able to travel in parallel entry with sudden diameter change, 89mm entry and 300 mm boiler header
- Able to enter and exit without any detached parts
- Able to traverse up to 4m long
- Capable in capturing 360° view of inner wall image with its location.
- Portable

3 Robot Design

Two major challenges in designing the robot are the size requirement and image capturing requirement. In order to fulfill the size and pipe accessibility requirement, the robot is designed to be a compact caterpillar type with track-based locomotion. Lifting mechanism and camera gimbal mechanism are designed and incorporated at the front side of the robot to ensure the image capturing can be carried out as required. Figure 1 shows the overall system of robot in pipe.

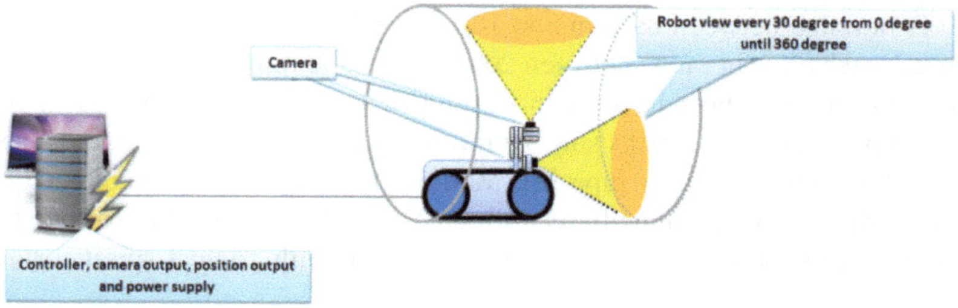

Fig. 1. Overall system of robot in pipe

This robot has 4 degree of freedom (DoF) which is forward and backward movement, turning movement, lifting and rotating. The robot model and fabricated prototype are shown in Figure 2.

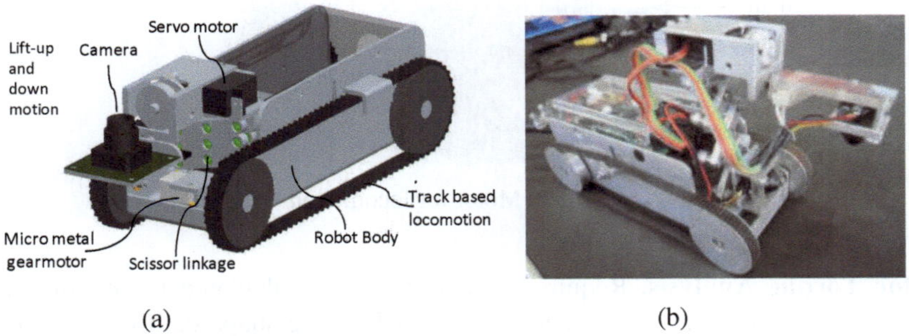

Fig. 2. (a) Prototype model (b) Fabricated prototype

Table 1 shows the mechanical structure of the robot. The overall robot dimensions are 192 mm (L) x 70 mm (W) x 70 mm (H). However, its height can be increased up to 92 percent of normal condition when necessary. The robot turning and built in camera features are discussed further in section 4. Track based feature assists the performance on holes and uneven surface conditions in pipe.

Table 1. Robot mechanical structure

Specifications	
Length	192 mm
Width	70 mm
Height	70 mm
Weight	670 g
Speed	0.10 m/s
Maximum Height	135 mm
No. of Degree of Freedom	4
Performance on holes	Able to travel on the holes back and forth
Performance on silence	No sound from motor
Turning Ability	Yes
Built in camera with 360° view	Yes

3.1 Transmission Mechanism

The movements are achieved by installing two Faulhaber Brushless DC motors Series 2232 BX4 CSD to obtain differential drive locomotion system. This motor is equipped with encoder as a sensor to locate the robot position. The small gear ratio of the motor, which is 4:1 provide the silence of the robot. Figure 2 shows the motor used for locomotion.

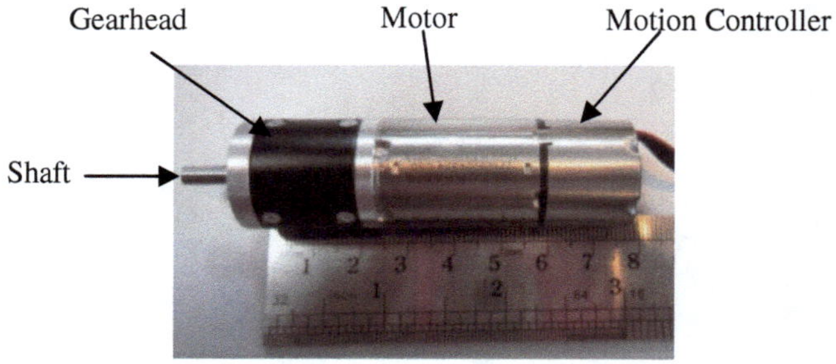

Fig. 3. Motor for locomotion

Motor Torque Analysis. Required torque has been calculated for motor selection process. Fig.3 shows boiler header with tapered nozzle shape configuration as robot environment. The robot needs to enter the entry nozzle, travel down the slope to get into boiler header pipe. Suitable motor required especially to travel up the slope when the inspection is over.

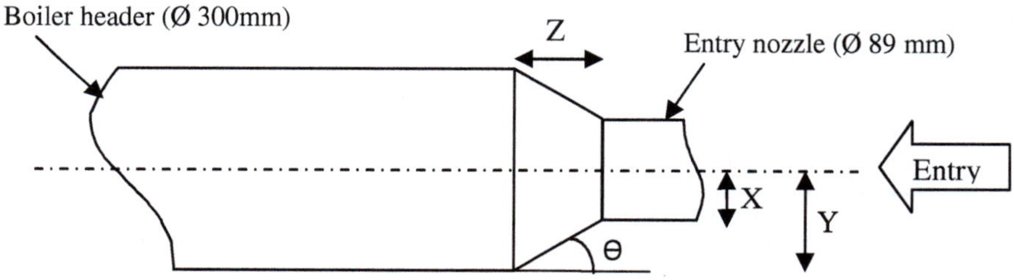

Fig. 4. Boiler header dimension

$$\theta = \tan^{-1}\frac{(x-y)}{z} \tag{1}$$

$$T = \frac{(a + g\sin\theta) \times M \times r}{N} \tag{2}$$

$$T_{req} = T \times eff_m \times eff_g \times SF \tag{3}$$

where a is the robot acceleration preliminary assumption of 0.1 ms^{-2}, Θ is the inclination angle. The slope inclination angle is 37°. M is the mass of robot and power cable, N is the number of motor used, and eff_m and eff_g are motor and gearhead efficiency respectively. Using safety factor of two, the calculation gives 276 mNm of torque requirement. Thus, brushless DC motor with a gear ratio of 4:1 is chosen to drive the robot. This motor has fulfilled the design requirement which is small, lightweight and has integrated encoder with 400 mNm continuous torque.

3.2 Lifting Mechanism

Due to the space limitation and high safety precaution, a single scissors linkage with a linear transmission is used for the lifting mechanism. This method is chosen due to the environment restriction which requires an expanding and subtracting mechanism. Figure 5 shows the structure of lifting mechanism Single scissors set is designed to reduce the parts of the robot and at the same time reduce the risk of fall apart parts. Linear transmission is achieved using a micro metal gear motor and leadscrew. Linear guide is used to give the stability to the scissors set especially when it is expanding. Figure 6 shows the lifting mechanism in closed configuration when entering the pipe entry and expanding to required height when necessary. The height of the camera can be varied from 70mm to 135mm from the ground. Figure 7 shows the motor used for lifting mechanism.

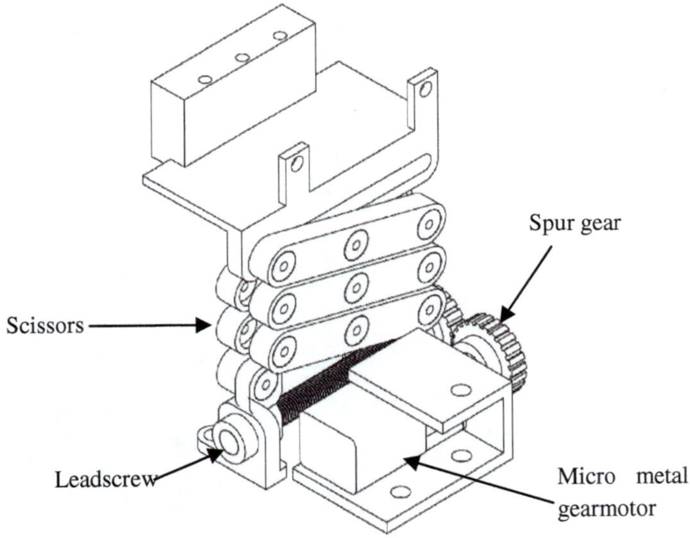

Fig. 5. Isometric view of lifting mechanism

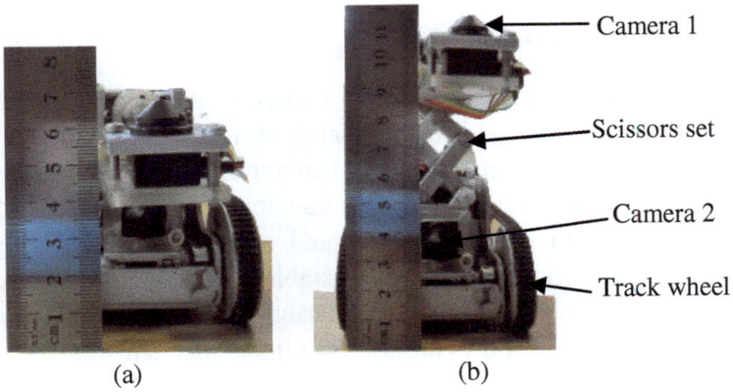

Fig. 6. Lifting mechanism (a) Closed configuration (b) Open configuration

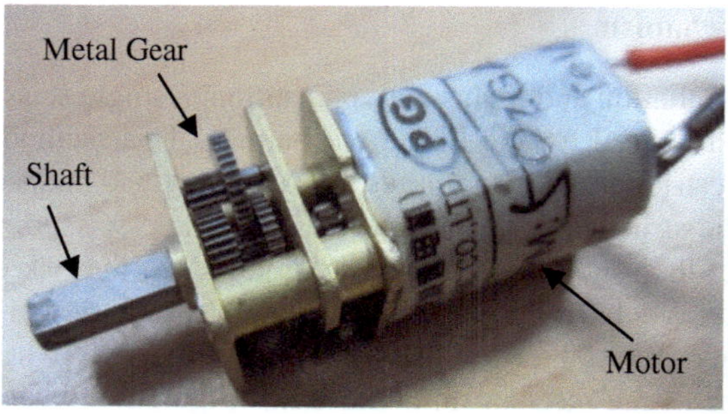

Fig. 7. Micro metal gearmotor (length, cross section size 20mmx20mm)

Bending Analysis. Pro Mechanica has been used to analyze bending behavior possibility on the 4mm diameter leadscrew used to lift the scissors. This is to ensure that the movement of the slider to lift the scissors is smooth. A significant displacement will cause immovable slider when sliding from one end to another. Figure 8 shows the analysis result of the bending analysis with 20N load. From the analysis, the maximum bending displacement is 0.00458mm can be justified as negligible. Moreover, the load carried by the leadscrew is less than 5N.

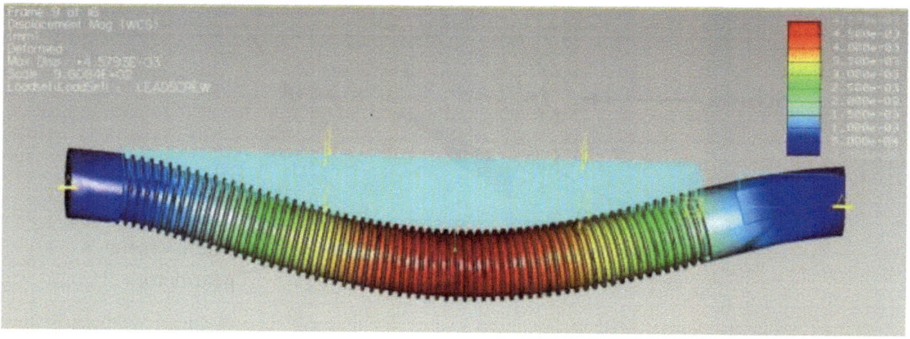

Fig. 8. Result of leadscrew bending analysis with 20N load

3.3 Camera Gimbal Mechanism

For rotation, a metal gear servo motor is used as the camera rotation actuator as it has position control ability. However, a servo motor can only rotates up to 180°. Due to space limitation restriction, this servo motor is connected with two spur gears with gear ratio of 2 to get a full rotation of camera. In order to get a whole image of inner surface this camera needs to be located outside the robot body. Extra support is needed to ensure that the camera is stable and will not fall apart from the robot. Therefore, a unique camera support is designed in addition to the support given by the scissors set. Figure 9b shows the camera gimbal mechanism.

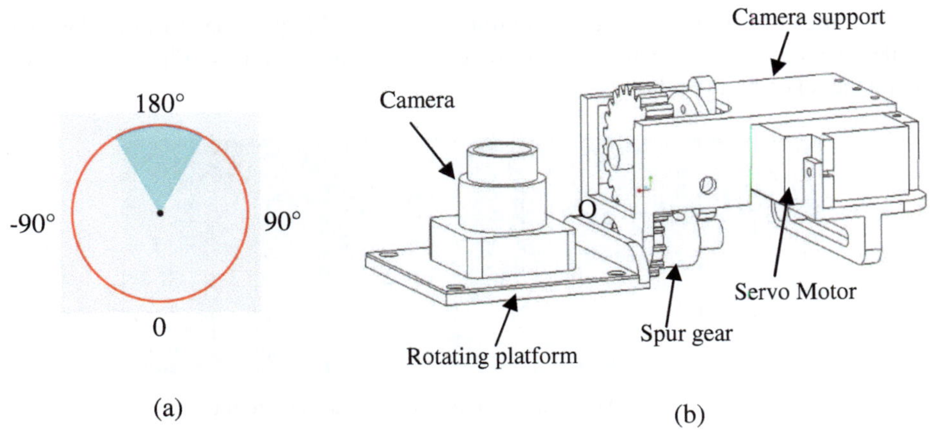

Fig. 9. a) Camera angle and direction b) Camera gimbal mechanism

Camera Stablity Analysis. As defined earlier in Section 3.3, the camera needs to be outside of the robot body. A stability analysis is performed to define the optimum position of camera on the robot. This is to ensure that the single set scissor linkage acting as pivot in Figure 10 is stable.

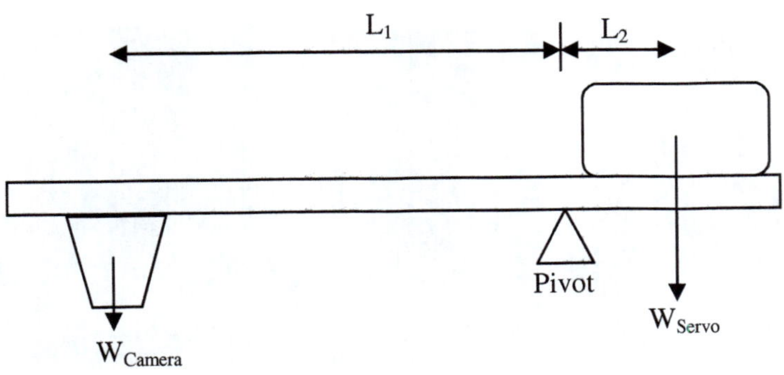

Fig. 10. Free body diagram of Camera gimbal mechanism

$$L_1 = \frac{W_{servo} \cdot L_2}{W_{camera}}, L_1 + L_2 > 30 \qquad (4)$$

where W_{servo} and W_{camera} are weight of servo and camera respectively, L_1 and L_2 are the length between pivot and camera, pivot and servo respectively. The condition of total length L_1 and L_2 is to ensure that the camera is outside of the robot body.

4 Experimental Result

4.1 Automated Image Acquisition Testing

This testing is carried out to get the optimum angle that give sufficient image quality. The camera is facing downward as starting position and rotated automatically at specific angle. From the testing, the optimum angle is 30° per image frame. Twelve frames are required to provide full coverage of inner surface image. Figure 11 shows the image taken by the robot.

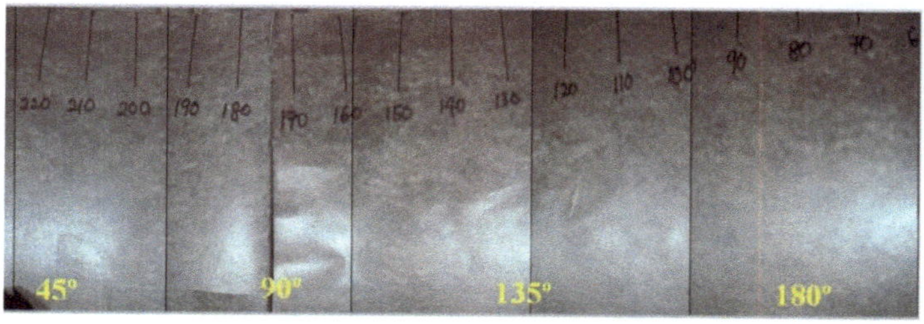

Fig. 11. Stitched image of inner surface taken by robot

4.2 Camera Rotation Testing

This experiment is carried out to test the stability and ability of camera gimbal mechanism to provide full rotation in its highest position. From the testing, the rotation angle can cater for 360 ° and can be precisely controlled. Figure 12 shows the camera position sequence to complete full rotation.

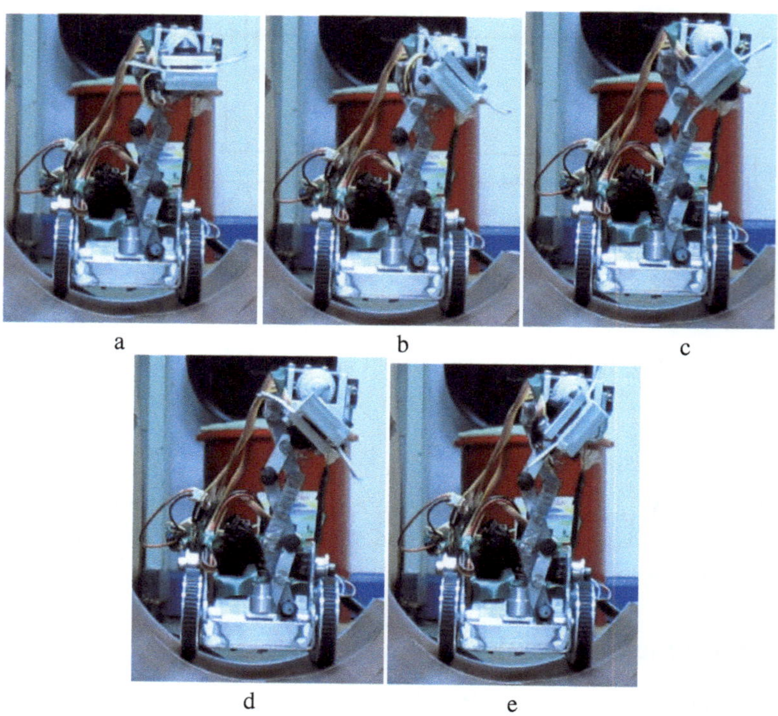

Fig. 12. Camera rotation sequence (a) 180° (b) 135° (c) 45° (d) -45° (e) -135°

4.3 Turning Testing

This testing is conducted to investigate the turning ability of the robot without any detached part. It is carried out on a steel surface pipeline with uneven surface condition. From the experiment, the track based transmission is able to perform differential drive locomotion without any detached part. Figure 13 shows the sequence of the robot in turning position.

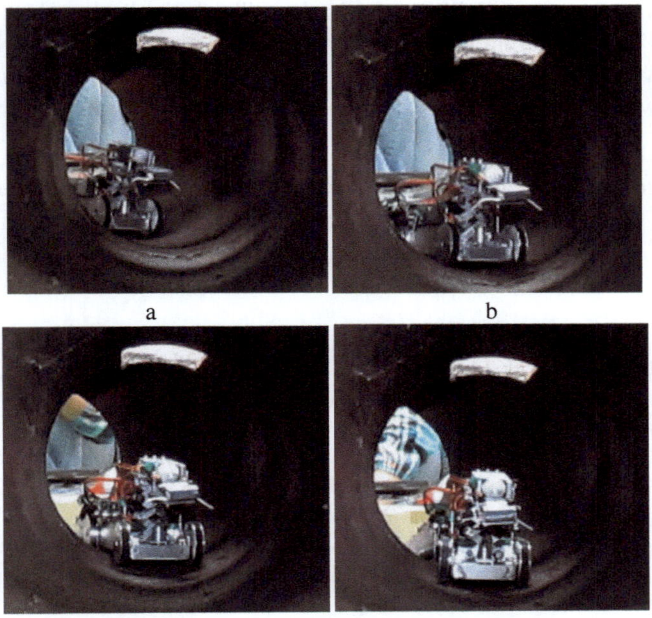

a b c d

Fig. 13. Turning ability sequence (a) Robot enters corner (b) Robot negotiating corner (c) Robot moves straight after corner (d) Robot enters straight path after corner

4.4 Slope Inclination Testing

This testing is carried out to assess the ability to travel on slope. The inclination angle tested is 40° which is higher than slope in boiler header. Figure 14 shows the sequence of the robot travel up the inclination slope. From the experiment, this robot manages to travel up the slope. This shows that the selected motor is suitable for the robot to travel inside the boiler header.

a b c d

Fig. 14. Slope Inclination Testing (a) 40 °ram (b) robot started to climb (c) robot is about to enter pipe (d) robot completes its task

5 Conclusion

A working prototype proved that the mechanisms and the robot is able to adapt drastic change in pipe diameter and acquire the image of internal pipe surface. Image can be viewed in 360° and location of the image taken can be obtained and stamped on the image. Adaptability of drastic change in diameter by the robot has been proven by subtraction and retraction of lifting mechanism. Tracked-wheel enables the robot to travel across uneven terrain such as inclined slope without getting stuck. The small size with overall dimension of 192 mm (L) x 70 mm (W) x 70 mm (H) and lightweight with only 0.67 kg makes the robot a portable and practical device.

Acknowledgement. The authors would like to thank Tenaga Nasional Berhad (TNB) and Ministry of Higher Education Malaysia (MOHE) for supporting this research.

References

1. Baharuddin, M.Z., Saad, J.M., Anuar, A., Ismail, I.N., Hassan Basri, N.M., Roslin, N.S., Kaja Mohideen, S.S., Abdul Jalal, M.F., Mohamed Sahari, K.S.: Robot for Boiler Header Inspection "LS-01". Procedia Engineering 41, 1483–1489 (2012)
2. Hoshina, M., Mashimo, T., Toyama, S.: Development of Spherical Ultrasonic Motor as a Camera Actuator for Pipe Inspection Robot. In: IEEE/RSJ International Conference on Intelligent Robots and Systems, pp. 2379–2384 (2009)
3. Brunete, A., Hernando, M., Gambao, E.: Modular Multiconfigurable Architecture for Low Diameter Pipe Inspection Microrobots. In: Proceeding of IEEE International Conference on Robotics and Automation, pp. 490–495 (2005)
4. Baba, T., Kameyama, Y., Kamegawa, T., Gofuku, A.: A snake robot propelling inside of a pipe with helical rolling motion. In: Proceeding of SICE Annual Conference, pp. 2319–2325 (2010)
5. Kuwada, A., Tsujino, K., Suzumori, K., Kanda, T.: Intelligent Actuators Realizing Snake-like Small Robot for Pipe Inspection. In: International Symposium on Micro-NanoMechatronics and Human Science, pp. 1–6 (2006)
6. Li, P., Ma, S., Li, B., Wang, Y.: Design of a Mobile Mechanism Possessing Driving Ability and Detecting Function for In-Pipe Inspection. In: IEEE International Conference on Robotics and Automation, pp. 3992–3997 (2008)
7. Choi, J.Y., Lim, H., Yi, B.-J.: Semi-Automatic Pipeline Inspection Robot Systems. In: SICE-ICASE International Joint Conference, pp. 2266–2269 (2006)
8. Quan, Q., Ma, S.: A Modular Crawler-driven Robot: Mechanical Design and Preliminary Experiments. In: IEEE/RSJ International Conference on Intelligent Robots and Systems, pp. 639–644 (2009)
9. Ohno, K., Morimura, S., Tadokoro, S., Koyanagi, E., Yoshida, T.: Semi-autonomous Control System of Rescue Crawler Robot Having Flippers for Getting Over Unknown-Steps. In: IEEE/RSJ International Conference on Intelligent Robos and Systems, pp. 3012–3018 (2007)
10. Kamimura, A., Kurokawa, H.: High-Step Climbing by a Crawler Robot DIR-2 - Ralization of Automatic Climbing Motionl. In: IEEE/RSJ International Conference on Intelligent Robots and Systems, pp. 618–624 (2009)

Author Index

MIX
Papier aus verantwortungsvollen Quellen
Paper from responsible sources
FSC® C105338

If you have any concerns about our products,
you can contact us on
ProductSafety@springernature.com

In case Publisher is established outside the EU,
the EU authorized representative is:
Springer Nature Customer Service Center GmbH
Europaplatz 3, 69115 Heidelberg, Germany

Printed by Libri Plureos GmbH
in Hamburg, Germany